Biomedical Materials for Multi-functional Applications

The series 'Biomedical Materials for Multi-functional Applications' is intended for the audience with an interest in biomaterials for various applications, including bio-implants, bio-instrumentation, biomechanics, and tissue engineering. The book titles under this series will bring different areas of biomedical engineering, bioengineering, materials science, mechanical engineering, and biosciences under one umbrella of biomedical materials. The books in this series will provide deep insights into the state-of-art of biomedical materials and deal with diverse and interdisciplinary aspects of biomaterials, ranging from basic science to host responses, as well as real-time applications. Some of the topics include but are not limited to: functional materials, smart materials, biomimetic materials, polymers, composites, ceramics, metals, alloys, and nanomaterials. This series will provide the most recent advances in biomedical materials, including design, synthesis, processing, and characterization. This series will accept both authored and edited volumes, including monographs, textbooks, reference works, and handbooks. This series aims to publish new advances in biomedical materials that have been made in the fabrication and development of biomaterials for different applications and the advancement of principles, theories, and designs. Each book title in the series will give a comprehensive overview of pioneering topics, and the readers will gain new knowledge and insights about the topic. As a collection, the series on 'Biomedical Materials for Multi-functional Applications' is anticipated to benefit a wide audience in academia, the research community, and industry.

Arnab Chanda · Gurpreet Singh

Soft Tissue Simulants

Springer

Arnab Chanda
Centre for Biomedical Engineering
Indian Institute of Technology Delhi
New Delhi, Delhi, India

Gurpreet Singh
Centre for Biomedical Engineering
Indian Institute of Technology Delhi
New Delhi, Delhi, India

ISSN 2731-9695 ISSN 2731-9709 (electronic)
Biomedical Materials for Multi-functional Applications
ISBN 978-981-97-3059-9 ISBN 978-981-97-3060-5 (eBook)
https://doi.org/10.1007/978-981-97-3060-5

This Springer imprint is published by the registered company Springer Nature Singapore Pte Ltd.
The registered company address is: 152 Beach Road, #21-01/04 Gateway East, Singapore 189721, Singapore

Preface

Soft tissue simulants, also known as synthetic tissues, have become a popular alternative to cadavers and animal tissues in various applications such as automotive crash testing, ballistic testing, and medical and surgical training. These simulants are designed to imitate the biomechanical properties of human soft tissues to provide accurate and realistic test results. In the past, cadavers and animal tissues were the primary materials used to simulate human soft tissues. However, cadaver tissues are not ideal for long-term use as their biomechanical properties change over time due to dehydration after death. Animal models, on the other hand, have different biomechanical properties that cannot be directly translated and compared to human tissues. To address these limitations, a wide range of soft tissue simulants have been developed, including gels, elastomers, and animal tissue surrogates. These simulants mimic the mechanical properties of human soft tissues more closely than traditional materials and are therefore more accurate and reliable. However, there has been a lack of technology transfer in this field, which has hindered the widespread adoption of these simulants. To address this issue, the book *Soft Tissue Simulants* aims to provide a comprehensive overview of the different soft tissue simulants, their characterization, and applications in various domains. The book covers different biofidelic simulants and includes a chapter on each tissue simulant, discussing its current trends and future challenges. By providing this information, the book aims to facilitate the adoption of more accurate and reliable soft tissue simulants in various industries.

The book is divided into 13 chapters, each of which covers a wide range of topics related to individual soft tissue simulants. Chapter 1 introduces the various types of tissues and their functions, along with the challenges associated with using cadaveric tissues. Chapter 2 focuses on the need for simulants and compares traditional tissues with soft tissue simulants. The chapter also highlights the hyperelastic modeling of soft tissue simulants. Skin, which is the largest tissue covering the human body, is discussed in Chap. 3. This chapter provides a detailed description of skin structure, the mechanical properties of skin, and the need and applications of different materials for skin simulants. Chapter 4 covers the different types of muscle tissues and discusses the fabrication of muscle tissue simulants. Chapter 5 explores the traditional testing of brain tissues, the need for brain tissue simulants, and the various application areas of brain tissue simulants. Chapters 6–8 focus on lung tissue, heart tissue, and breast tissue, individually. These chapters discuss the traditional testing of cadaveric tissues, the need for tissue simulants, compositions and materials for fabricating tissue simulants, and the application areas of these particular tissues, such as lung tissue, heart tissue, and breast tissue. Similarly, Chaps. 9–11 focus on abdominal tissues, where Chap. 9 corresponds to liver tissue, Chap. 10 corresponds to stomach tissue, and Chap. 11 corresponds to gallbladder tissue. Chapter 12 discusses the different types of arteries, their significance, and the need for arterial tissue simulants. The chapter also elaborates on the need for novel arterial tissue simulants and their applications in different areas. Finally, Chap. 13 presents the state-of-the-art of ballistic tissue simulants. Starting with ballistic simulants for brain tissue, this chapter also discusses the skull simulants and skin simulants, highlighting their proficiency in ballistic testing.

Overall, this book is a valuable resource for anyone interested in soft tissue simulants and their various applications. The book is written in a clear and concise manner, making it accessible to both experts and novices in the field. These materials have the potential to substitute live tissues for medical interventions, without requiring extensive biosafety approvals. Additionally, the growing community of academic researchers working on soft tissue biomechanics could benefit greatly from this knowledge, as these biomaterials can replace real tissues in testing, thereby reducing the cost and time involved in lab work. This book is expected to serve as a comprehensive reference textbook for research in several fields, including tissue engineering and biomedical engineering, medical simulation, biomechanics, finite element modeling of biological systems, biomaterials, biotechnology, implant and medical device development, and healthcare wearables.

New Delhi, India
March 2024

Arnab Chanda
arnab.chanda@cbme.iitd.ac.in

Gurpreet Singh
gurpreet.singh@cbme.iitd.ac.in

Contents

About the Authors

Dr. Arnab Chanda is Assistant Professor in the Centre for Biomedical Engineering, IIT Delhi, and a joint faculty at the Department of Biomedical Engineering, AIIMS, Delhi. He is also the founder of a start-up company Biofit Technologies LLC, USA. He earned his Ph.D. from the University of Alabama, USA and post-doctoral degree at the University of Pittsburgh, USA. Dr. Chanda is an expert in the fabrication and mechanical characterization of tissue mimics, and has previously developed artificial surrogates for human skin, muscles, brain, artery, and plantar fascia, and tested them at both lab and clinical settings. These experimental models have been used extensively for surgical training and have led to development of several healthcare products and startups.

Dr. Chanda is a recipient of the Early Career Faculty Research Award and Young Researcher awards from ASME and MHRD and AO Spine International-Singapore. He has 10+ US and Indian Patent applications and has authored four books and 100+ articles in reputed international journals. Dr. Chanda is in the list of top 2% scientists worldwide as per Scopus Elsevier 2023 database. He is also serving as the editor for a book series published by Springer Nature, editorial board member for Nature Scientific Reports, coordinator of the "Executive Programme in Healthcare Entrepreneurship and Management", and guest editor for several international journals of repute.

Gurpreet Singh is a Ph.D. scholar in the Centre for Biomedical Engineering, Indian Institute of Technology Delhi, India. He is a recipient of the most prestigious Ph.D. fellowship in India, the Prime Minister's Research Fellowship (PMRF), in the May 2021 cycle. His research interests are soft tissue mechanics, artificial tissues, biomimetics, and computational biomechanics. He is working on the computational modelling of diabetic foot and its ulceration progression. He is also working in developing artificial human tissue surrogates for injury and disease modelling. Previously, much of his work has been on improving the surface characteristics and bioactivity of metallic biomaterials, with research interests, including surface engineering, materials science, biomaterials, and non-conventional machining processes. He worked on the surface modification of metallic biomaterials using electro-discharge machining, where he studied the bioactivity of modified surfaces in terms of wear resistance, corrosion resistance, and other biological responses. He has authored/edited two books and contributed 50+ research papers and book chapters to leading international journals and conferences. He also serves as a reviewer for prominent international journals of repute.

Chapter 1
Introduction

Human tissues can be broadly classified into two main categories: hard tissues and soft tissues. These classifications are based on the structural and functional characteristics of the tissues. Both are two distinct categories of biological tissues, each with unique properties and functions. The main differences between hard and soft tissues lie in their composition, structure, and mechanical properties. Hard tissues are characterized by the presence of a mineralized extracellular matrix, mainly hydroxyapatite crystals (calcium and phosphate compounds), making them rigid and stiff. Hard tissues have a well-defined structure and are often organized into layers or lamellae. Hard tissues provide structural support, protection, and rigidity to the body. Bones, for example, support the body, protect internal organs, and serve as a reservoir for minerals. Hard tissues are typically found in skeletal structures, such as bones and teeth. Hard tissues are rigid and have high compressive strength. They are designed to withstand mechanical forces and provide stability to the body [1–3].

On the other hand, soft tissues are primarily composed of cells and extracellular matrix. The extracellular matrix is often rich in proteins like collagen and elastin, providing flexibility and elasticity to the tissue. Soft tissues are generally pliable and flexible. They do not have a rigid structure and are deformable to a certain extent. Soft tissues consist of several layers, with distinct compositions for each layer [4]. Epithelial tissue, connective tissue, muscle tissue, and neural tissue are thought to be the four types of typical tissues. From a mechanical perspective, connective tissues are often regarded as the most significant for soft tissue mechanical research [5, 6]. Collagen, reticular, and elastic fibers are the three kinds of fibers that make up the extracellular matrix, in addition to the base material. Since collagen fibers account for the majority of mechanical behavior, they are often seen as being more significant than other types, especially due to their size. The collagen fibers are connected to the elastic fibers, which are mostly made of elastin and exhibit a purely elastic behavior. These fibers are primarily responsible for the elastic properties of soft tissues. Soft tissues serve various functions, such as providing support, connecting and enclosing organs, facilitating movement, and participating in metabolic processes

A. Chanda and G. Singh, *Soft Tissue Simulants*, Biomedical Materials for Multi-functional Applications, https://doi.org/10.1007/978-981-97-3060-5_1

[7]. Soft tissues encompass a diverse range of types, each with distinct functions and properties. The major types of soft tissues include muscle tissue, connective tissue, nervous tissue, blood vessels, skin, nerves, and a wide range of functional tissues. The functional tissues include the brain, tongue, tonsils, heart, esophagus, small intestine, colon, kidney, liver, gallbladder, lungs, breast, stomach, spleen, urinary bladder, uterus, and vagina. These soft tissues collectively contribute to the structure, function, and overall well-being of the human body, and exhibit characteristics like elasticity and compliance. They can undergo deformation without breaking and can return to their original shape after stretching or compression. It's important to note that within each category, some further subtypes and variations fulfill specific roles in different regions of the body [8].

1.1 Functions and Types of Different Soft Tissues

- **Muscle Tissue**: The muscle tissues are responsible for the movement and locomotion of the human body. The three types of muscle tissues are skeletal muscle, smooth muscle, and cardiac muscle. The skeletal muscle is attached to bones and is responsible for voluntary movement. The smooth muscles form the walls of the internal organs (e.g., stomach, intestines) and control involuntary movement, such as peristalsis. The human heart consists of cardiac muscle and takes care of the involuntary contraction to pump blood [9].
- **Connective Tissue**: The connective tissues provide structural support, bind tissues and organs together, and play a role in metabolic processes. The different types of connective tissues are loose connective tissue, dense connective tissue, adipose tissue, blood, bone, and cartilage. The loose connective tissue binds organs together and provides support. The dense connective tissue consists of regular (tendons, ligaments) and irregular (dermis of skin) connective tissues, which provide strength and support. Another important connective tissue is adipose tissue, which stores fat, insulates, and cushions organs [9].
- **Vascular Tissue**: The vascular tissue is categorized as arteries, veins, and capillaries. The arteries carry oxygenated blood away from the heart, veins carry deoxygenated blood toward the heart, and capillaries facilitate the exchange of gases and nutrients between blood and tissues [10].
- **Skin**: Skin tissue is a type of biological tissue that forms the outermost protective covering of the body. It is the largest organ of the integumentary system and serves several crucial functions, such as protection, sensation, regulation of temperature, and synthesis of Vitamin D. The skin is composed of multiple layers (epidermis, dermis, hypodermis), each with distinct functions and characteristics [11, 12].
- **Functional Tissues**: The human body's functioning organs span a broad variety of soft tissues, both in length and width (Fig. 1.1). The brain, tongue, tonsils, salivary glands, esophagus, lungs, heart, breast, liver, gallbladder, stomach, small intestine, colon, spleen, pancreas, kidney, urine bladder, uterus, and vagina are among the primary functional soft tissues [8].

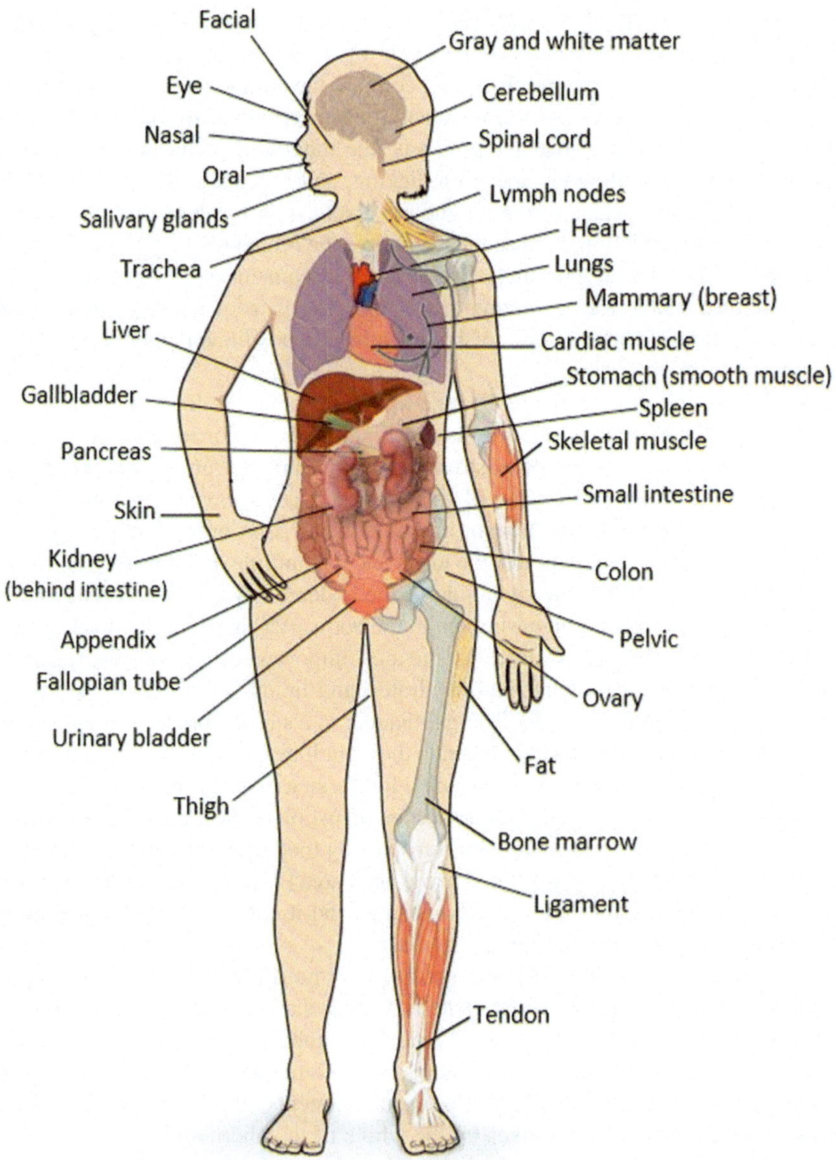

Fig. 1.1 Soft tissues located across the human body [10]

One of the wrinkly organs that can be found inside the cranial cavity is the brain tissue, which is the central nervous system of the human body. The intracranial cerebrum, the cerebellum, and the spinal cord are the components that make up this structure since it is responsible for transmitting information to the body [13, 14]. As a functional organ of the upper airway, the tongue plays an important role in the processes of swallowing, speaking, licking, and breathing. The taste buds that are located on the surface of the tongue are responsible for detecting the flavor of food, and the information is subsequently sent to the brain [15–17]. In order to avoid infections in the throat and lungs, tonsils are a component of our immune system. They are responsible for preventing germs and viruses from entering our body and preventing illness. These are the fresh lumps that are positioned on both sides of the mouth.

The lungs are a pair of large, sponge-like organs that are located in the thorax, lateral to the heart and superior to the diaphragm. The major function of the lungs is to provide oxygen to the bloodstream and, in return, to remove carbon dioxide from the bloodstream (via the use of very small air sacs known as alveoli). Located in the thoracic region of the human body, the heart is an organ that pumps blood throughout the body. In relation to the lungs and the midline of the body, it is situated in the frontal region. On the right side of the abdomen is where you will find the liver, which is the second biggest organ in the body after the skin. The liver tissue is responsible for a number of functions, the most important of which is the regulation of the metabolism of proteins, carbohydrates, and lipids, as well as the storage of minerals and vitamins [18, 19]. The gallbladder is a slender organ in the form of a pear and is attached immediately beneath the right lobe of the liver. Through the use of various lipids, it is primarily responsible for the storage and release of bile into the duodenum, which is necessary for the normal digestion of the meal [20]. A muscular tube, the esophagus is responsible for transporting food and water to the stomach by way of the pharynx, which is located in the neck area [21, 22]. The cardiac sphincter is located at the inferior end of the esophagus, and it is responsible for sealing the esophagus and capturing food in the stomach.

At the same time as it serves as a repository for nutrition, the stomach is a significant organ of the gastrointestinal system. A muscular sac that may be found in the upper belly, between the esophagus and the duodenum [23], is a part of the digestive tract. The duodenum, jejunum, and ileum are all components of the lower gastrointestinal tract, whereas the small intestine is a component of this tract [24]. It is situated underneath the stomach and takes up the whole of the abdominal cavity during its position. In the same way that a hose is coiled, the small intestine has the ability to facilitate digestion and the absorption of nutrients [25]. It is somewhat inferior to the stomach and coils around the top and lateral border of the small intestine. The colon, which is also referred to as the big intestine, is located in this position. A flattened, oval-shaped piece of soft tissue, the spleen may be found on the upper left side of the abdomen, lateral to the stomach. The spleen is equipped with a sizable blood reserve that assists in the process of filtering huge quantities of blood [26]. A place in the abdominal cavity that is immediately inferior and posterior to the stomach is where the tissue of the pancreas is located. It has the appearance of a lumpy snake, with its "head" connected to the duodenum and its "tail" pointing to the left wall of the abdominal cavity. Within the small intestine, digestive enzymes are secreted by the pancreas. These enzymes are responsible for the chemical breakdown of carbohydrates. Located along the posterior wall of the abdominal cavity, the kidney is a bean-shaped tissue that is used for renal function. In addition to being situated posterior to the peritoneum, the kidneys are in touch with the muscles that are placed in the back. The kidneys are shielded from harm by a thin layer of adipose tissue, which also helps to hold them in place and avoids bodily damage. The urinary bladder is a hollow organ that is situated in the middle of the body, close to the base of the pelvis, and is responsible for collecting urine and eventually excreting it [27]. The vaginal tissue is a tube that is both elastic and muscular, and it serves as a delivery route for the transport of maternal blood during birthing as well as a pathway for menstrual blood [28, 29]. Around 7.5 cm along the front wall and 9 cm along the posterior wall are the dimensions of the vaginal tissue. After passing through the cervix, the vaginal tissue is attached to the tissue of the uterus as well as the fallopian tubes [30, 31].

1.2 Cadaveric Tissues for Research Purposes

While experiments with cadaveric tissues provide valuable insights into human anatomy, physiology, and medical education, there are several ethical, practical, and technical issues associated with such experiments. Some of these issues include

- **Ethical Concerns**: Using cadaveric tissues raises ethical questions about the treatment of human remains and the necessity of obtaining informed consent from donors or their families. Cultural and religious beliefs regarding the treatment of human remains may conflict with the use of cadaveric tissues for research [32, 33].
- **Storage and Handling**: Maintaining the structural integrity of cadaveric tissues is challenging. Proper preservation methods are required to prevent decomposition and maintain tissue quality for experimentation [10].
- **Limited Availability**: The availability of cadaveric tissues is finite, and access may be restricted, limiting the number and scope of experiments that can be conducted [10].
- **Biological Variation**: Each cadaver is unique, and there can be considerable biological variation among donors. This variability may affect the generalizability of experimental results [10].
- **Transmission of Infections**: Despite careful handling and preservation, there is always a small risk of infection transmission from cadaveric tissues, especially if proper precautions are not taken [10].
- **Costs and Logistics: Expense**: Maintaining a cadaveric tissue program is costly due to storage facilities, preservation methods, and associated logistics [10].

- **Legal and Regulatory Compliance**: There are legal and regulatory frameworks governing the use of cadaveric tissues, and institutions must adhere to these guidelines to ensure ethical and legal compliance [10].

Despite these challenges, cadaveric tissues remain an invaluable resource for medical education, surgical training, and anatomical research. Proper ethical considerations, adherence to regulations, and advancements in alternative methods can help address some of the associated issues. However, in recent advances soft tissue simulants are in trend to mimic the properties of the natural soft tissues [34–36].

1.3 Soft Tissue Simulants

Soft tissue simulants are materials designed to replicate the mechanical and physical properties of human soft tissues (see Fig. 1.2). These simulants play a crucial role in various scientific and engineering applications, providing researchers, engineers, and medical professionals with a way to test and validate devices, techniques, and procedures in a controlled and reproducible manner. Soft tissue simulants are used in fields such as biomechanics, medical device development, surgical training, and imaging technology testing. Some of the commonly used materials for soft tissue simulants are gelatin, silicone rubbers, polyurethane foams, hydrogels, thermoplastic elastomers, and agar-gelatin mixtures [10, 37–39].

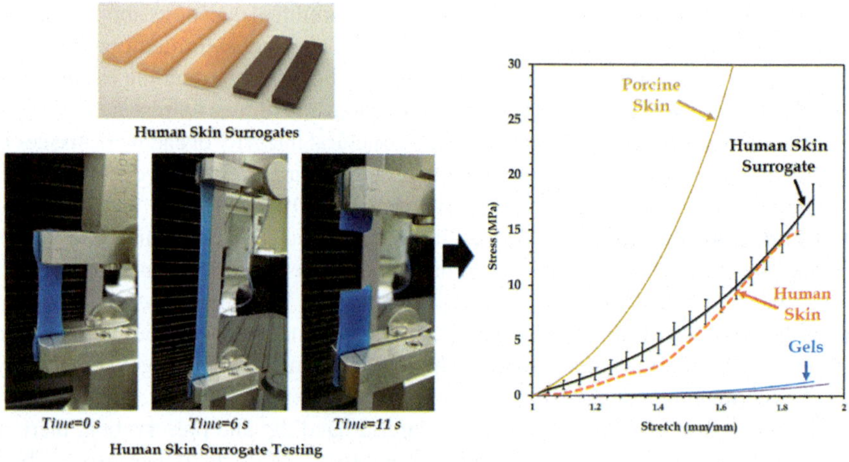

Fig. 1.2 Soft tissue simulants [40]

- **Gelatin**: Gelatin is derived from collagen, a protein found in animal connective tissues. Gelatin-based simulants can closely mimic the mechanical properties of certain soft tissues, making them suitable for applications in surgery and medical device testing [41].
- **Silicone Rubbers**: Silicone rubbers are synthetic polymers with elastomeric properties. They are often used to simulate skin and soft tissue due to their flexibility, elasticity, and durability. Different formulations can mimic various soft tissue characteristics and are used in medical training, device testing, and prosthetics development [41].
- **Polyurethane Foams**: Polyurethane foams are polymers that can be engineered to have a wide range of mechanical properties. These foams can be tailored to replicate the density and elasticity of specific soft tissues. They are commonly used in crash test dummies to simulate human tissue during impact studies [41].
- **Ballistics Gel**: It is used in forensic and ballistics testing to simulate the effects of bullets or other projectiles on soft tissues. Ballistic gel mimics the density and viscosity of human tissue [41].
- **Hydrogels**: Hydrogels are water-absorbing polymers with a high water content, making them similar to biological tissues. Hydrogels can be designed to mimic the mechanical and rheological properties of soft tissues. They are widely used in tissue engineering and regenerative medicine for simulating biological tissues [41].
- **Thermoplastic Elastomers (TPE)**: TPEs are a class of polymers that combine the properties of rubber and thermoplastics. TPEs can be used to simulate the elastic behavior of soft tissues. They are suitable for applications requiring repeated deformations. Used in medical and surgical simulations, as well as for developing realistic tissue models for training purposes, thermoplastic elastomers can be customized to mimic the elasticity and texture of different soft tissues [41].
- **Agar-Gelatin Mixtures**: Mixtures of agar and gelatin can provide a balance between the mechanical properties of the two materials. These mixtures are often used to simulate brain tissue for surgical training and medical device testing. The choice of soft tissue simulant depends on the specific application and the tissue properties that need to be replicated. Also, it can simulate the mechanical properties of tissues, especially in applications related to incisions, suturing, and dissection [41].
- **Natural Tissue Simulants**: Sometimes, animal tissues or synthetic materials are combined to create more realistic simulants, termed natural tissue simulants. Properties: Customizable to simulate specific tissue types [41].

The choice of soft tissue simulant depends on the specific application and the properties that need to be replicated. For medical training, it's crucial to have simulants that closely resemble the characteristics of human soft tissues to provide a realistic and effective learning experience. Researchers and developers often work on improving the accuracy and complexity of soft tissue simulants to enhance their utility in various fields [42, 43]. In addition, soft tissues or their surrogates are considered as nonlinear hyperelastic material. Numerous hyperelastic models, including the Ogden, Yeoh,

Mooney-Rivlin, Neo-Hookean, and polynomial models, have been proposed in the literature [40, 44, 45]. The frameworks of hyperelastic models are almost identical, with the only variations being in the complexity or quantity of material factors included. When it comes to material behavior under quasi-static loadings, or within a very narrow range of strain rates, the hyperelastic models accurately predict the results.

1.4 Summary

In summary, soft tissues are essential for maintaining the structural integrity and proper functioning of the body. They often work in conjunction with other tissues and organs to perform complex physiological processes. The study of soft tissues is critical in various medical and scientific fields, including anatomy, physiology, pathology, and biomedical engineering. Researchers and medical professionals continually explore new technologies and techniques for imaging, analyzing, and treating soft tissues to advance our understanding of health and disease [8, 46–49].

Skin, vascular tissues, neurological tissues, skeletal muscle, cartilage, and ligaments are examples of soft tissues that might lose their normal functioning as a result of pathological conditions, cancer, or trauma. These types of soft tissues need either repair or improvement in order to ensure that the injured tissue continues to function properly [50]. The functioning, on the other hand, needs to take into account not just the mechanical qualities of the tissue but also its chemical and biological features. Moreover, the new material must have qualities that are completely biocompatible with the existing ones. Polymers have been the subject of substantial research as a tissue replacement biomaterial for the purpose of constructing soft tissues. These soft tissues include skin, heart valves, nerves, arteries, liver, pancreas, bladder, and other soft tissues [50–55].

Some of the natural polymers that are used extensively by researchers to imitate the mechanical properties of natural soft tissues include collagen, alginate, elastin, chitosan, polylactide (PLA), poly-glycerol-sebacate (PGS), poly (lactic-co-glycolic acid) (PLGA), and polyurethane (PU) [56–60]. These polymers are examples of natural polymers. In addition to polymers, hydrogels are well-known for their contributions to tissue engineering due to the fact that they are able to imitate the actual mechanical characteristics of human soft tissues [61–63].

Performing experimental tests on cadaveric tissues for the purpose of ex vivo testing is challenging due to a number of factors, including ethical constraints, the accessibility and availability of tissue samples, the reproducibility of findings across samples, and the presence of variable boundary conditions [64]. Soft tissue simulants mimic the mechanical and physical properties of human tissues for various purposes, including medical device testing, surgical training, and biomechanical research. They provide consistent and reproducible properties, reducing variability in test results compared to cadaveric tissues, which may vary in quality and condition.

References

1. Makode S, Singh G, Chanda A (2021) Development of novel anisotropic skin simulants. Phys Scr 96:125019. https://doi.org/10.1088/1402-4896/AC2EFD
2. Chanda A, McClain S (2019) Mechanical modeling of healthy and diseased calcaneal fat pad surrogates. Biomimetics 4:1. https://doi.org/10.3390/BIOMIMETICS4010001
3. Chanda A, Unnikrishnan V, Lackey K, Robbins J (2020) Biofidelic conductive soft tissue surrogates. Int J Polym Mater Polym Biomater 69:127–135. https://doi.org/10.1080/00914037.2018.1552856
4. Singh G, Gupta V, Chanda A (2022) Artificial skin with varying biomechanical properties. Mater Today Proc 62:3162–3166. https://doi.org/10.1016/J.MATPR.2022.03.433
5. Natali AN, Carniel EL, Gregersen H (2009) Biomechanical behaviour of esophageal tissues: material and structural configuration, experimental data and constitutive analysis. Med Eng Phys 31:1056–1062. https://doi.org/10.1016/J.MEDENGPHY.2009.07.003
6. Epstein E, Munderloh N (1975) Isolation and characterization of CNBr peptides of human (alpha 1 (III))3 collagen and tissue distribution of (alpha 1 (I))2 alpha 2 and (alpha 1 (III))3 collagens. J Biol Chem 250:9304–9312. https://doi.org/10.1016/s0021-9258(19)40644-3
7. Gupta V, Singh G, Chanda A (2022) Development and testing of skin grafts models with varying slit orientations. Mater Today Proc 62:3462–3467. https://doi.org/10.1016/J.MATPR.2022.04.282
8. Chanda A, Singh G (2023) Introduction to human tissues. Mater Horizons From Nat to Nanomater 1–12. https://doi.org/10.1007/978-981-99-2225-3_1/COVER
9. Chanda A, Singh G (2023) Muscles and connective tissues. Mater Horizons Nat Nanomater 25–32. https://doi.org/10.1007/978-981-99-2225-3_3/COVER
10. Singh G, Chanda A (2021) Mechanical properties of whole-body soft human tissues: a review. Biomed Mater 16:062004
11. McGrath JA, Uitto J (2010) Anatomy and organization of human skin. In: Rook's textbook of dermatology, vol 1. Wiley-Blackwell, Oxford, pp 1–53. https://doi.org/10.1002/9781444317633.ch3
12. Chanda A, Unnikrishnan V, Lackey K (2017) Biofidelic conductive synthetic skin composites. In: 32nd technical conference of the American Society for composites 2017, vol 1. DEStech Publications Inc., pp 400–408. https://doi.org/10.12783/asc2017/15197
13. Budday S, Sommer G, Birkl C, Langkammer C, Haybaeck J, Kohnert J et al (2017) Mechanical characterization of human brain tissue. Acta Biomater 48:319–340. https://doi.org/10.1016/j.actbio.2016.10.036
14. Budday S, Sarem M, Starck L, Sommer G, Pfefferle J, Phunchago N et al (2020) Towards microstructure-informed material models for human brain tissue. Acta Biomater 104:53–65. https://doi.org/10.1016/j.actbio.2019.12.030
15. Hermant N, Perrier P, Payan Y (2017) Human tongue biomechanical modeling. In: Biomechanics of living organs: hyperelastic constitutive laws for finite element modeling. Elsevier Inc., pp 395–411. https://doi.org/10.1016/B978-0-12-804009-6.00019-5
16. Kajee Y, Pelteret J-PV, Reddy BD (2013) The biomechanics of the human tongue. Int J Numer Method Biomed Eng 29:492–514. https://doi.org/10.1002/cnm.2531
17. Payan Y, Bettega G, Raphaël B (1998) A biomechanical model of the human tongue and its clinical implications. Lecture notes in computer science (including subseries Lecture notes in artificial intelligence and Lecture notes in bioinformatics), vol 1496. Springer, pp 688–695. https://doi.org/10.1007/bfb0056255
18. Kazemnejad S (2009) Hepatic tissue engineering using scaffolds: state of the art. Avicenna J Med Biotechnol 1:135–145
19. Lorenzini S, Andreone P (2007) Stem cell therapy for human liver cirrhosis: a cautious analysis of the results. Stem Cells 25:2383–2384. https://doi.org/10.1634/stemcells.2007-0056
20. Li WG, Hill NA, Ogden RW, Smythe A, Majeed AW, Bird N et al (2013) Anisotropic behaviour of human gallbladder walls. J Mech Behav Biomed Mater 20:363–375. https://doi.org/10.1016/j.jmbbm.2013.02.015

21. Mir M, Ali MN, Ansari U, Sami J (2016) Structure and motility of the esophagus from a mechanical perspective. Esophagus 13:8–16. https://doi.org/10.1007/s10388-015-0497-1
22. Hajhosseini P, Takalloozadeh M (2019) An isotropic hyperelastic model of esophagus tissue layers along with three-dimensional simulation of esophageal peristaltic behavior. J Bioeng Res 1:12–27. https://doi.org/10.22034/jbr.2019.189018.1009
23. Brandstaeter S, Fuchs SL, Aydin RC, Cyron CJ (2019) Mechanics of the stomach: a review of an emerging field of biomechanics. GAMM-Mitteilungen 42. https://doi.org/10.1002/gamm.201900001
24. Bellini C, Glass P, Sitti M, Di Martino ES (2011) Biaxial mechanical modeling of the small intestine. J Mech Behav Biomed Mater 4:1727–1740. https://doi.org/10.1016/j.jmbbm.2011.05.030
25. Miyasaka EA, Okawada M, Utter B, Mustafa-Maria H, Luntz J, Brei D et al (2010) Application of distractive forces to the small intestine: defining safe limits. J Surg Res 163:169–175. https://doi.org/10.1016/j.jss.2010.03.060
26. Kemper AR, Santago AC, Stitzel JD, Sparks JL, Duma SM (2012) Biomechanical response of human spleen in tensile loading. J Biomech 45:348–355. https://doi.org/10.1016/j.jbiomech.2011.10.022
27. Barnes SC, Shepherd DET, Espino DM, Bryan RT (2015) Frequency dependent viscoelastic properties of porcine bladder. J Mech Behav Biomed Mater 42:168–176. https://doi.org/10.1016/j.jmbbm.2014.11.017
28. Chanda A, Unnikrishnan V, Roy S, Richter HE (2015) Computational modeling of the female pelvic support structures and organs to understand the mechanism of pelvic organ prolapse: a review. Appl Mech Rev 67. https://doi.org/10.1115/1.4030967
29. Ashton-Miller JA, Delancey JOL (2007) Functional anatomy of the female pelvic floor. Ann N Y Acad Sci 1101. https://doi.org/10.1196/annals.1389.034
30. Li X, Kruger JA, Chung J-H, Nash MP, Nielsen PMF (2008) Modelling childbirth: comparing athlete and non-athlete pelvic floor mechanics. Springer, Berlin, Heidelberg, pp 750–757. https://doi.org/10.1007/978-3-540-85990-1_90
31. Baah-Dwomoh A, McGuire J, Tan T, De Vita R (2016) Mechanical properties of female reproductive organs and supporting connective tissues: a review of the current state of knowledge. Appl Mech Rev 68. https://doi.org/10.1115/1.4034442
32. Gupta V, Chanda A (2022) Biomechanics of skin grafts: effect of pattern size, spacing and orientation. Eng Res Express 4:015006. https://doi.org/10.1088/2631-8695/AC48CB
33. Gupta V, Chanda A (2022) Expansion potential of skin grafts with alternating slit based auxetic incisions. Forces Mech 7:100092. https://doi.org/10.1016/J.FINMEC.2022.100092
34. Singh G, Chanda A (2022) Biomechanical modeling of progressive wound healing: a computational study. Biomed Eng Adv 4:100055. https://doi.org/10.1016/J.BEA.2022.100055
35. Gupta V, Chanda A (2023) Expansion potential of novel skin grafts simulants with I-shaped auxetic incisions. Biomed Eng Adv 5:100071. https://doi.org/10.1016/J.BEA.2023.100071
36. Singh G, Chanda A (2023) Development and mechanical characterization of artificial surrogates for brain tissues. Biomed Eng Adv 5:100084. https://doi.org/10.1016/J.BEA.2023.100084
37. Gupta V, Singh G, Chanda A (2023) Development of hierarchical auxetic skin graft simulants with high expansion potential. Biomed Eng Adv 5:100087. https://doi.org/10.1016/J.BEA.2023.100087
38. Gupta V, Singla R, Singh G, Chanda A (2023) Development of soft composite based anisotropic synthetic skin for biomechanical testing. Fibers 11:55. https://doi.org/10.3390/FIB11060055
39. Gupta V, Gupta S, Chanda A (2023) Development of an ultra-low-cost planar biaxial tester for soft tissue characterization. Biomed Phys Eng Express 9:025011. https://doi.org/10.1088/2057-1976/ACB940
40. Chanda A (2018) Biomechanical modeling of human skin tissue surrogates. Biomimetics 3:18. https://doi.org/10.3390/BIOMIMETICS3030018
41. Singh G, Chanda A (2021) Mechanical properties of whole-body soft human tissues: a review. Biomed Mater 16:062004. https://doi.org/10.1088/1748-605X/AC2B7A

42. Gupta V, Singh G, Chanda A (2023) Development of novel hierarchical designs for skin graft simulants with high expansion potential. Biomed Phys Eng Express 9:035024. https://doi.org/10.1088/2057-1976/ACC661
43. Gupta V, Singh G, Gupta S, Chanda A (2023) Expansion potential of auxetic prosthetic skin grafts: a review. Eng Res Express 5:022003. https://doi.org/10.1088/2631-8695/ACCFE5
44. Singh G, Chanda A (2023) Biofidelic gallbladder tissue surrogates. Adv Mater Process Technol. https://doi.org/10.1080/2374068X.2023.2198835
45. Singh G, Chanda A (2023) Development and biomechanical testing of artificial surrogates for vaginal tissue. Adv Mater Process Technol. https://doi.org/10.1080/2374068X.2023.2198837
46. Chanda A, Callaway C (2018) Tissue anisotropy modeling using soft composite materials. Appl Bionics Biomech 2018. https://doi.org/10.1155/2018/4838157
47. Chanda A, Unnikrishnan V, Flynn Z, Lackey K (2017) Experimental study on tissue phantoms to understand the effect of injury and suturing on human skin mechanical properties. Proc Inst Mech Eng Part H J Eng Med 231:80–91. https://doi.org/10.1177/0954411916679438/ASSET/IMAGES/LARGE/10.1177_0954411916679438-FIG16.JPEG
48. Chanda A, Singh G (2023) Applications, challenges, and future opportunities. In: Materials horizons: from nature to nanomaterials, pp 85–92. https://doi.org/10.1007/978-981-99-2225-3_8/COVER
49. Chanda A, Singh G (2023) Tissues in functional organs—low stiffness. In: Materials horizons: from nature to nanomaterials, pp 33–48. https://doi.org/10.1007/978-981-99-2225-3_4/COVER
50. Dhandayuthapani B, Yoshida Y, Maekawa T, Kumar DS (2011) Polymeric scaffolds in tissue engineering application: a review. Int J Polym Sci. https://doi.org/10.1155/2011/290602
51. Mayer J, Karamuk E, Akaike T, Wintermantel E (2000) Matrices for tissue engineering-scaffold structure for a bioartificial liver support system. J Control Release (Elsevier) 64:81–90. https://doi.org/10.1016/S0168-3659(99)00136-4
52. Oberpenning F, Meng J, Yoo JJ, Atala A (1999) De novo reconstitution of a functional mammalian urinary bladder by tissue engineering. Nat Biotechnol 17:149–155. https://doi.org/10.1038/6146
53. Tziampazis E, Sambanis A (1995) Tissue engineering of a bioartificial pancreas: modeling the cell environment and device function. Biotechnol Prog 11:115–126. https://doi.org/10.1021/bp00032a001
54. Pabari A, Lloyd-Hughes H, Seifalian AM, Mosahebi A (2014) Nerve conduits for peripheral nerve surgery. Plast Reconstr Surg 133:1420–1430. https://doi.org/10.1097/PRS.0000000000000226
55. Germain L, Auger FA, Grandbois E, Guignard R, Giasson M, Boisjoly H et al (1999) Reconstructed human cornea produced in vitro by tissue engineering. Pathobiology 67:140–147. https://doi.org/10.1159/000028064
56. Bressan E, Favero V, Gardin C, Ferroni L, Iacobellis L, Favero L et al (2011) Biopolymers for hard and soft engineered tissues: application in odontoiatric and plastic surgery field. Polymers (Basel) 3:509–526. https://doi.org/10.3390/polym3010509
57. Del Bakhshayesh AR, Asadi N, Alihemmati A, Tayefi Nasrabadi H, Montaseri A, Davaran S et al (2019) An overview of advanced biocompatible and biomimetic materials for creation of replacement structures in the musculoskeletal systems: focusing on cartilage tissue engineering. J Biol Eng 13:1–21. https://doi.org/10.1186/s13036-019-0209-9
58. West JL (2020) Biomaterials for cardiovascular tissue engineering. Biomater Sci (Elsevier) 1389–1397. https://doi.org/10.1016/b978-0-12-816137-1.00086-6
59. Nguyen PK, Baek K, Deng F, Criscione JD, Tuan RS, Kuo CK (2020) Tendon tissue-engineering scaffolds. Biomater Sci (Elsevier) 1351–1371. https://doi.org/10.1016/b978-0-12-816137-1.00084-2
60. Siddiqui N, Asawa S, Birru B, Baadhe R, Rao S (2018) PCL-based composite scaffold matrices for tissue engineering applications. Mol Biotechnol 60:506–532. https://doi.org/10.1007/s12033-018-0084-5

61. Talebian S, Mehrali M, Taebnia N, Pennisi CP, Kadumudi FB, Foroughi J et al (2019) Self-healing hydrogels: the next paradigm shift in tissue engineering? Adv Sci 6. https://doi.org/10.1002/advs.201801664
62. Wang Y, Adokoh CK, Narain R (2018) Recent development and biomedical applications of self-healing hydrogels. Expert Opin Drug Deliv 15:77–91. https://doi.org/10.1080/17425247.2017.1360865
63. Mondal S, Das S, Nandi AK (2020) A review on recent advances in polymer and peptide hydrogels. Soft Matter 16:1404–1454. https://doi.org/10.1039/c9sm02127b
64. Ottensmeyer MP, Kerdok AE, Howe RD, Dawson SL (2004) The effects of testing environment on the viscoelastic properties of soft tissues. Lecture notes in computer science (including subseries Lecture notes in artificial intelligence and Lecture notes in bioinformatics), vol 3078, pp 9–18. https://doi.org/10.1007/978-3-540-25968-8_2

Chapter 2
Emerging Need for Simulants

2.1 Need for Soft Tissue Simulants

Soft tissue simulants are materials designed to mimic the properties of human soft tissues, and their emerging need is particularly evident in various fields such as medicine, surgery, prosthetics, and medical device testing. The emerging need for soft tissue simulants is driven by several factors including medical training and simulation, medical device testing and development, forensic science, cosmetic and aesthetic procedures, prosthetics and robotics, and research and development [1–4]. The emerging need for soft tissue simulants is driven by advancements in materials science, 3D printing, and a growing emphasis on realistic and safe training and testing environments in various fields. These simulants contribute to more effective medical practices, improved product development, and enhanced research outcomes.

Soft tissue simulants are crucial for training surgeons and medical professionals [2]. They provide a realistic environment for practicing surgical procedures without the need for live patients. Simulants help train healthcare professionals in various medical procedures, such as injections, wound care, and ultrasound examinations [5]. Soft tissue simulants are essential for testing the safety and efficacy of medical implants and devices before they are used in humans. This ensures that the devices interact appropriately with surrounding tissues. Researchers use soft tissue simulants to study the biomechanics of tissues and develop devices that can interact with the human body more effectively. Soft tissue simulants are employed in testing the biocompatibility of medical devices. This ensures that devices interact appropriately with human tissues and do not cause adverse reactions. Simulants help test the effectiveness and safety of medical devices that involve needle insertions, such as injection systems and catheters [6–8].

Soft tissue simulants play a role in training for robotic-assisted surgeries, providing a platform for practicing precise movements and interactions with soft tissues. Simulants aid in the development and testing of instruments used in minimally invasive surgeries, ensuring they perform optimally within soft tissue environments. Soft

A. Chanda and G. Singh, *Soft Tissue Simulants*, Biomedical Materials for Multi-functional Applications, https://doi.org/10.1007/978-981-97-3060-5_2

tissue simulants are used to simulate human tissue in ballistics testing [1, 9, 10]. This helps forensic experts understand the effects of bullets and other projectiles on the human body. Simulants aid in recreating crime scenes for forensic investigations, allowing experts to analyze the trajectory and impact of injuries. Soft tissue simulants are also widely used in cosmetic and aesthetic procedures. They are utilized in the training of practitioners for cosmetic procedures such as injections and dermal fillers. This ensures that practitioners are proficient in these techniques before performing them on patients. Companies developing cosmetic and aesthetic products use soft tissue simulants to test the safety and effectiveness of their products. Soft tissue simulants are used in the development and testing of prosthetic limbs and organs, ensuring that these devices replicate the feel and function of real human tissues [11–13]. Soft tissue simulants are employed to study how robots interact with the human body, facilitating the development of more human-friendly robotic systems. Simulants are used to test and develop devices such as braces and supports that interact with soft tissues. In research and development, soft tissue simulants are used to simulate human tissues for drug delivery studies, helping researchers understand how medications diffuse and interact within the body. Soft tissue simulants support various biomedical research studies by providing a controlled and reproducible environment for experimentation [14–16].

2.2 Comparison of Soft Tissue Simulants with Traditional Tissues

Soft tissue simulants are materials designed to mimic the mechanical and physical properties of biological tissues, often used in medical training, research, and product testing. Traditional tissues, on the other hand, refer to actual biological tissues obtained from humans or animals.

2.2.1 Source and Availability

Soft tissue simulants can be manufactured and produced on demand. There are no ethical concerns or limitations on availability with consistent quality and quantity. The traditional tissues are usually obtained from human or animal donors. There are limited availability and ethical considerations with variability in quality and quantity [3, 4].

2.2.2 Ethical Considerations

Soft tissue simulants provide an ethical alternative to using animal or human tissues for training or testing purposes. The use of traditional tissues raises ethical concerns, especially when sourced from animals or cadavers [4, 6].

2.2.3 Consistency and Reproducibility

Soft tissue simulants offer consistency and reproducibility in terms of their mechanical properties. They can be manufactured with specific properties, allowing for standardized testing and training. Human and animal tissues may exhibit natural variations, making it challenging to ensure consistency in experiments or training scenarios [4].

2.2.4 Customization

Soft tissue simulants can be customized to replicate specific tissues or pathologies, allowing for targeted training or testing scenarios. While traditional tissues can offer realism, they may not always provide the level of customization available with simulants [4].

2.2.5 Durability

Soft tissue simulants are often more durable and have a longer shelf life compared to traditional tissues, which may degrade over time. Human or animal tissues are subject to decomposition, requiring special storage conditions and limited usability [4, 17].

2.2.6 Cost

Soft tissue simulants can be more cost-effective in the long run due to their durability and reproducibility. The collection, preservation, and storage of human or animal tissues can be expensive, and their use is often limited by availability [4, 18–20].

2.2.7 *Realism*

While soft tissue simulants aim to replicate the mechanical properties of tissues, they may not always perfectly mimic the complexity and feel of real tissues. Human or animal tissues offer a high level of realism, especially in terms of texture and response to surgical instruments [4].

2.2.8 *Regulatory Compliance*

Soft tissue simulants may need to meet certain regulatory standards for medical training and testing. The use of human tissues may involve strict ethical and regulatory considerations, including consent and proper sourcing [14, 15].

2.2.9 *Transportation and Storage*

Soft tissue simulants are easy to transport and store under various conditions. They are less susceptible to temperature and environmental variations. These kinds of tissues require specific storage conditions, such as refrigeration or freezing. Also, the transport and storage of traditional tissues can be logistically challenging [4].

In summary, soft tissue simulants and traditional tissues each have their advantages and disadvantages. Soft tissue simulants offer advantages in terms of ethical considerations, consistency, customizability, and cost-effectiveness. However, traditional tissues may still be preferred in certain contexts where the specific properties of natural tissues are crucial, despite the associated challenges [21–23]. The choice between soft tissue simulants and traditional tissues depends on the specific requirements of the application, balancing factors such as cost, availability, customization needs, and ethical considerations. Soft tissue simulants offer advantages in terms of reproducibility and customization, while traditional tissues provide a higher level of realism.

2.3 Hyperelastic Modeling of Soft Tissue Simulants

Hyperelastic modeling is a mathematical approach used to describe the nonlinear, time-independent mechanical behavior of materials, including soft tissues. Soft tissue simulants, designed to mimic the mechanical properties of human tissues, often utilize hyperelastic models to accurately represent their deformation characteristics under various loading conditions that occur during various medical procedures and interventions. These models are especially valuable in fields such as biomechanics, medical device design, and surgical simulation [24, 25].

2.3.1 *Constitutive Equations*

Hyperelastic models are based on constitutive equations that relate the stress and strain in a material. These equations capture the nonlinear, time-independent, and large deformation behavior of soft tissues. Some of the commonly used hyperelastic models for soft tissues include the Neo-Hookean model, Mooney-Rivlin model, Ogden model, etc. [26, 27].

2.3.2 *Material Parameters*

The accuracy of hyperelastic models relies on determining material parameters that characterize the specific mechanical properties of the soft tissue simulant. Material parameters may be obtained through experimental testing, such as uniaxial or biaxial tension tests, compression tests, and indentation tests [28, 29].

2.3.3 *Experimental Validation*

It is essential to validate hyperelastic models by comparing their predictions with experimental data. Experimental validation helps ensure that the chosen model accurately represents the mechanical behavior of the soft tissue simulant. Strain, stress, and deformation data obtained from physical testing are used to refine and validate the model parameters [30–32].

2.3.4 *FEM Simulation*

Hyperelastic models are often implemented in finite element simulations to predict the mechanical response of soft tissue simulants in complex scenarios. Finite element analysis allows for the simulation of various loading conditions, such as stretching, compression, and shearing, providing insights into the behavior of soft tissues under different circumstances [33–36].

2.3.5 Applications in Medical Simulation

Hyperelastic modeling is particularly relevant in medical simulation, where soft tissue simulants are used to replicate the mechanical properties of organs and tissues. Surgical simulators leverage hyperelastic models to provide realistic feedback to trainees, allowing them to practice procedures in a virtual environment [9, 37, 38].

2.3.6 Biomechanics Studies

Hyperelastic modeling supports biomechanics studies by enabling researchers to analyze the mechanical response of soft tissues to external forces. Understanding the hyperelastic behavior of tissues is crucial for designing medical implants, prosthetics, and devices that interact appropriately with the human body [17, 24, 27, 37, 39, 40].

2.3.7 Material Characterization Techniques

Techniques such as inverse finite element analysis and optimization algorithms are employed to determine material parameters by comparing model predictions with experimental data. In addition, some hyperelastic models can be extended to include viscoelastic components to account for time-dependent behavior. This is important when studying the dynamic response of soft tissues [41–43].

In summary, hyperelastic modeling is a valuable tool for accurately representing the mechanical behavior of soft tissue simulants. This approach enhances the realism of simulations and contributes to advancements in medical training, device development, and biomechanics research. Experimental validation and careful consideration of material parameters are essential steps in ensuring the accuracy and reliability of hyperelastic models for soft tissue simulants [28, 33, 44]. The choice of a hyperelastic model depends on the specific soft tissue being simulated and the level of accuracy required. More complex models may better capture the material behavior but may also require more parameters, which may be challenging to determine experimentally. In practical applications, finite element analysis (FEA) is often used in conjunction with hyperelastic models to simulate and analyze the deformation of soft tissues in response to external forces or deformations [19, 43, 45]. Experimentally derived material properties are used to calibrate the hyperelastic models for accurate simulations.

References

1. Jussila J, Leppäniemi A, Paronen M, Kulomäki E (2005) Ballistic skin simulant. Forensic Sci Int 150:63–71. https://doi.org/10.1016/J.FORSCIINT.2004.06.039
2. Payne T, Mitchell S, Bibb R, Waters M (2015) Development of novel synthetic muscle tissues for sports impact surrogates. J Mech Behav Biomed Mater 41:357–374. https://doi.org/10.1016/j.jmbbm.2014.08.011
3. Makode S, Singh G, Chanda A (2021) Development of novel anisotropic skin simulants. Phys Scr 96:125019. https://doi.org/10.1088/1402-4896/AC2EFD
4. Singh G, Chanda A (2021) Mechanical properties of whole-body soft human tissues: a review. Biomed Mater 16:062004. https://doi.org/10.1088/1748-605X/AC2B7A
5. Chanda A, Unnikrishnan V, Lackey K, Robbins J (2020) Biofidelic conductive soft tissue surrogates. Int J Polym Mater Polym Biomater 69:127–135. https://doi.org/10.1080/00914037.2018.1552856
6. Singh G, Gupta V, Chanda A (2022) Artificial skin with varying biomechanical properties. Mater Today Proc 62:3162–3166. https://doi.org/10.1016/J.MATPR.2022.03.433
7. Gupta V, Singh G, Chanda A (2022) Development and testing of skin grafts models with varying slit orientations. Mater Today Proc 62:3462–3467. https://doi.org/10.1016/J.MATPR.2022.04.282
8. Gupta V, Chanda A (2022) Biomechanics of skin grafts: effect of pattern size, spacing and orientation. Eng Res Express 4:015006. https://doi.org/10.1088/2631-8695/AC48CB
9. Bracq A, Haugou G, Delille R, Lauro F, Roth S, Mauzac O (2017) Experimental study of the strain rate dependence of a synthetic gel for ballistic blunt trauma assessment. J Mech Behav Biomed Mater 72:138–147. https://doi.org/10.1016/J.JMBBM.2017.04.027
10. Casem DT, Dwivedi AK, Mrozek RA, Lenhart JL (2014) Compression response of a thermoplastic elastomer gel tissue surrogate over a range of strain-rates. Int J Solids Struct 51:2037–2046. https://doi.org/10.1016/J.IJSOLSTR.2013.12.028
11. Chen RK, Jin Y, Wensman J, Shih A (2016) Additive manufacturing of custom orthoses and prostheses—a review. Addit Manuf 12:77–89. https://doi.org/10.1016/J.ADDMA.2016.04.002
12. Gupta V, Singh G, Gupta S, Chanda A (2023) Expansion potential of auxetic prosthetic skin grafts: a review. Eng Res Express 5:022003. https://doi.org/10.1088/2631-8695/ACCFE5
13. Chhikara K, Sidhu SS, Gupta S, Saharawat S, Kataria C, Chanda A (2023) Development and effectiveness testing of a novel 3D-printed multi-material orthosis in nurses with plantar foot pain. Prosthes 5:73–87. https://doi.org/10.3390/PROSTHESIS5010006
14. Gupta V, Chanda A (2022) Expansion potential of skin grafts with alternating slit based auxetic incisions. Forces Mech 7:100092. https://doi.org/10.1016/J.FINMEC.2022.100092
15. Chhikara K, Singh G, Gupta S, Chanda A (2022) Progress of additive manufacturing in fabrication of foot orthoses for diabetic patients: a review. Ann 3D Print Med 8:100085. https://doi.org/10.1016/J.STLM.2022.100085
16. Singh G, Chanda A (2022) Biomechanical modeling of progressive wound healing: a computational study. Biomed Eng Adv 4:100055. https://doi.org/10.1016/J.BEA.2022.100055
17. Singh G, Chanda A (2023) Biofidelic gallbladder tissue surrogates. Adv Mater Process Technol. https://doi.org/10.1080/2374068X.2023.2198835
18. Chanda A, Singh G (2023) Muscles and connective tissues. In: Materials horizons: from nature to nanomaterials, pp 25–32. https://doi.org/10.1007/978-981-99-2225-3_3/COVER
19. Chanda A, Singh G (2023) Tissues in functional organs—low stiffness. In: Materials horizons: from nature to nanomaterials, pp 33–48. https://doi.org/10.1007/978-981-99-2225-3_4/COVER
20. Chanda A, Singh G (2023) Introduction to human tissues. In: Materials horizons: from nature to nanomaterials, pp 1–12. https://doi.org/10.1007/978-981-99-2225-3_1/COVER
21. Gupta V, Chanda A (2023) Expansion potential of novel skin grafts simulants with I-shaped auxetic incisions. Biomed Eng Adv 5:100071. https://doi.org/10.1016/J.BEA.2023.100071
22. Singh G, Chanda A (2023) Development and mechanical characterization of artificial surrogates for brain tissues. Biomed Eng Adv 5:100084. https://doi.org/10.1016/J.BEA.2023.100084

23. Gupta V, Singh G, Chanda A (2023) Development of hierarchical auxetic skin graft simulants with high expansion potential. Biomed Eng Adv 5:100087. https://doi.org/10.1016/J.BEA.2023.100087
24. Chanda A, Singh G (2023) Hyperelastic models for anisotropic tissue characterization. In: Materials horizons: from nature to nanomaterials, pp 73–83. https://doi.org/10.1007/978-981-99-2225-3_7/COVER
25. Hajhosseini P, Takalloozadeh M (2019) An isotropic hyperelastic model of esophagus tissue layers along with three-dimensional simulation of esophageal peristaltic behavior. J Bioeng Res 1:12–27. https://doi.org/10.22034/jbr.2019.189018.1009
26. Almeida ES, Spilker RL (1998) Finite element formulations for hyperelastic transversely isotropic biphasic soft tissues. Comput Methods Appl Mech Eng 151:513–538. https://doi.org/10.1016/S0045-7825(97)82246-3
27. Concha F, Sarabia-Vallejos M, Hurtado DE (2018) Micromechanical model of lung parenchyma hyperelasticity. J Mech Phys Solids 112:126–144. https://doi.org/10.1016/j.jmps.2017.11.021
28. Petre M, Erdemir A, Panoskaltsis VP, Spirka TA, Cavanagh PR (2013) Optimization of nonlinear hyperelastic coefficients for foot tissues using a magnetic resonance imaging deformation experiment. J Biomech Eng 135. https://doi.org/10.1115/1.4023695/370866
29. Lapeer RJ, Gasson PD, Karri V (2010) Simulating plastic surgery: from human skin tensile tests, through hyperelastic finite element models to real-time haptics. Prog Biophys Mol Biol 103:208–216. https://doi.org/10.1016/j.pbiomolbio.2010.09.013
30. Zhu Z, Jiang C, Jiang H (2019) A visco-hyperelastic model of brain tissue incorporating both tension/compression asymmetry and volume compressibility. Acta Mech 230:2125–2135. https://doi.org/10.1007/s00707-019-02383-1
31. Eghbali R, Narooei K (2022) A hyperelastic-damage model to study the anisotropic mechanical behavior of coral-hydrogel bio-composite. J Mech Behav Biomed Mater 126:105054. https://doi.org/10.1016/J.JMBBM.2021.105054
32. Martins PALS, Natal Jorge RM, Ferreira AJM (2006) A comparative study of several material models for prediction of hyperelastic properties: application to silicone-rubber and soft tissues. Strain 42:135–147. https://doi.org/10.1111/j.1475-1305.2006.00257.x
33. Sasson A, Patchornik S, Eliasy R, Robinson D, Haj-Ali R (2012) Hyperelastic mechanical behavior of chitosan hydrogels for nucleus pulposus replacement—experimental testing and constitutive modeling. J Mech Behav Biomed Mater 8:143–153. https://doi.org/10.1016/J.JMBBM.2011.12.008
34. Weiss JA, Maker BN, Govindjee S (1996) Finite element implementation of incompressible, transversely isotropic hyperelasticity. Comput Methods Appl Mech Eng 135:107–128. https://doi.org/10.1016/0045-7825(96)01035-3
35. May-Newman K, Lam C, Yin FCP (2009) A hyperelastic constitutive law for aortic valve tissue. J Biomech Eng 131. https://doi.org/10.1115/1.3127261/398327
36. Groves RB, Coulman SA, Birchall JC, Evans SL (2013) An anisotropic, hyperelastic model for skin: experimental measurements, finite element modelling and identification of parameters for human and murine skin. J Mech Behav Biomed Mater 18:167–180. https://doi.org/10.1016/j.jmbbm.2012.10.021
37. Hermant N, Perrier P, Payan Y (2017) Human tongue biomechanical modeling. In: Biomechanics of living organs: hyperelastic constitutive laws for finite element modeling. Elsevier Inc., pp 395–411. https://doi.org/10.1016/B978-0-12-804009-6.00019-5
38. Valencia A, Burdiles P, Ignat M, Mura J, Bravo E, Rivera R et al (2013) Fluid structural analysis of human cerebral aneurysm using their own wall mechanical properties. Comput Math Methods Med 2013. https://doi.org/10.1155/2013/293128
39. Chanda A, Unnikrishnan V, Flynn Z, Lackey K (2017) Experimental study on tissue phantoms to understand the effect of injury and suturing on human skin mechanical properties. Proc Inst Mech Eng Part H J Eng Med 231:80–91. https://doi.org/10.1177/0954411916679438/ASSET/IMAGES/LARGE/10.1177_0954411916679438-FIG16.JPEG
40. Morales-Orcajo E, Souza TR, Bayod J, Barbosa de Las Casas E (2017) Non-linear finite element model to assess the effect of tendon forces on the foot-ankle complex. Med Eng Phys 49:71–78. https://doi.org/10.1016/J.MEDENGPHY.2017.07.010

41. Singh G, Chanda A (2023) Development and biomechanical testing of artificial surrogates for vaginal tissue. Adv Mater Process Technol. https://doi.org/10.1080/2374068X.2023.2198837
42. Singh G, Chanda A (2023) Development and biomechanical testing of human stomach tissue surrogates. In: Materials horizons: from nature to nanomaterials, Part F1471, pp 113–125. https://doi.org/10.1007/978-981-99-5064-5_7/COVER
43. Singh G, Chanda A (2023) Biofidelic tongue and tonsils tissue surrogates. In: Materials horizons: from nature to nanomaterials, Part F1471, pp 159–170. https://doi.org/10.1007/978-981-99-5064-5_10/COVER
44. Chanda A, Callaway C (2018) Tissue anisotropy modeling using soft composite materials. Appl Bionics Biomech 2018. https://doi.org/10.1155/2018/4838157
45. Gupta V, Singh G, Chanda A (2023) Modeling of metamaterial based incision patterns for generating high expansions in skin grafts. Clin Biomech 110:106118. https://doi.org/10.1016/J.CLINBIOMECH.2023.106118

Chapter 3
Skin Tissue Simulants

The skin, being the largest organ in terms of surface area, has many functions including protecting against the external environment, regulating body temperature, and conforming to the body's shape during movement. This tissue exhibits mechanical complexity due to its multi-layered structure, including diverse components in each layer. The mechanical characteristics of the skin are of great importance for many medicinal, cosmetic, and biomechanical applications. Applications in cosmetics include assessing the emollient properties and moisturizing effects of cosmetic products. Medical professionals use the biomechanical characteristics of the skin to get a deeper understanding of suturing methods, skin disease treatments, plastic surgery operations, and other related areas [1, 2]. The biomechanical characteristics of skin tissues are important for developing and creating synthetic materials that mimic human skin in real-time anthropomorphic devices, such as crash test dummies and surgical simulators. The material characteristics of the skin play a crucial role in the creation and analysis of computer models of the human body for biomechanical simulations, as well as their use in physical biomechanical models [3].

3.1 Skin Structure

The human skin is a material that exhibits anisotropic properties, and its properties vary depending on the direction in which they are measured. It is also non-homogeneous, and does not have a uniform composition throughout. Additionally, the skin is viscoelastic, meaning it displays both viscous and elastic behavior. The surface area of an adult's skin is around two square meters, which accounts for an average body mass of 5.5% [4–6]. The epidermis, dermis, and hypodermis are the three main layers comprising the intricate, multi-layered structure (Fig. 3.1) [7, 8]. Human skin has a thickness that varies between 0.3 and 2.6 mm in different parts

A. Chanda and G. Singh, *Soft Tissue Simulants*, Biomedical Materials for Multi-functional Applications, https://doi.org/10.1007/978-981-97-3060-5_3

of the body [9], with an average thickness of 2 mm [6]. The dermis layer consists mostly of fibrous proteins such as collagen, reticulin, and elastin. On the other hand, the epidermis layer is a thin covering that spans most of the human body. It measures between 0.07 and 0.12 mm in thickness, with the palms measuring 0.8 mm and the soles averaging 1.4 mm. The thickness of the dermis ranges from an average of 1–2 mm to as low as 0.6 mm in some regions [10]. It is commonly acknowledged that dermis thickness plays a crucial role in the mechanical properties of skin tissue [8]. The hypodermis, located under the dermis layer and without a distinct boundary, consists mostly of connective tissue and adipose lobules [11–13]. The hypodermis, the third layer of the skin, is connected to the fascia by a honeycomb structure. It functions as the body's fat storage region and has a strong attachment to the dermis, as well as a flexible and loose connection to other tissues, such as interior soft tissue.

The epidermis layer consists of the stratum corneum, stratum granulosum, stratum spinosum, and stratum basale as its sub-layers. The dermis, located under the epidermis layer and possessing an elastic support structure, plays a vital role in the mechanical properties of the skin tissue [8]. The main components of it consist of nerve endings, blood capillaries, hair follicles, sweat glands, and lymphatics. The dermis thickness is 0.7 mm on the eyelids and 1.5 mm on the back, abdomen, buttock, thigh, and dorsum of the foot, as determined in previous studies [10, 15]. The dermis exhibits variations in thickness throughout the body, often averaging 1–2 mm. The epidermis layer contains keratinocytes, cells, and cellular debris, while the dermis layer mostly consists of fibroblasts and fibrous proteins such as elastin, collagen fibers, and reticulin. Collagen fibers consist of collagen fibrils, which are compact

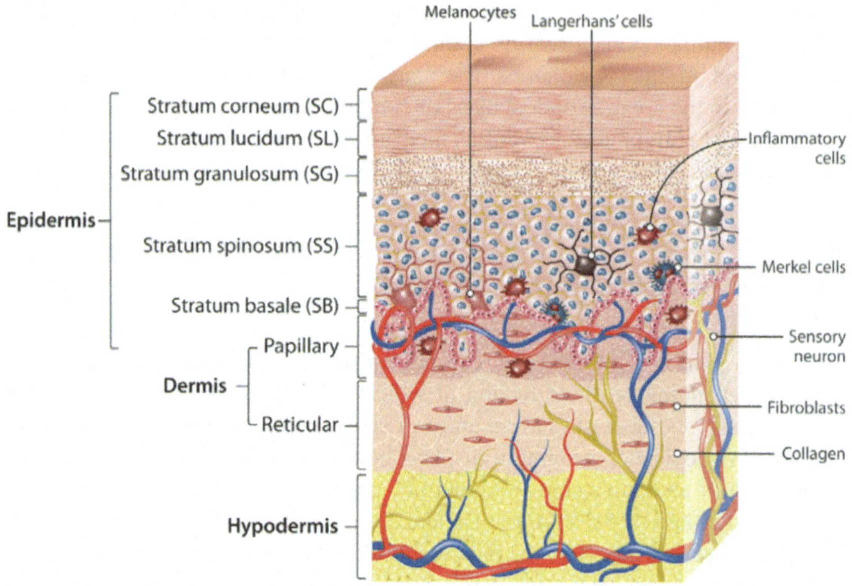

Fig. 3.1 The multilayer structure of human skin tissue [14]

and thick fibers measuring between 60 and 100 nm in size [16]. The dermis layer consists of two connective tissues: the papillary dermis, which is the thin top layer, and the reticular dermis, which is the thick bottom layer. These tissues are situated under the epidermal layer and above the subcutaneous layer.

3.2 Mechanical Properties of Skin

The mechanical qualities of the skin are associated with skin thickness, which is influenced by differences in dermal thickness and the amount of collagen and elastin present [17]. The hypodermis layer functions as an energy reservoir, regulates body temperature, and acts as a cushion to mitigate the impact of injuries on the underlying tissues and organs. The hypodermis layer provides a pliable link to other tissues and is connected to the fascia by a honeycomb structure (known as skin-internal soft tissue) [11–13]. The thickness of the hypodermis also varies depending on its location on the body. The hypodermis layer is thickest on the plantar surface of the feet, the palmar surface of the hands, and the buttocks region.

The skin's anisotropic thin membrane layers exhibited mechanical features such as resistance to friction and a reaction to lateral compression [18–20]. Because of its uneven morphology and nonlinear stress–strain correlation, the skin is constantly subjected to tension. Pre-stress lines, sometimes referred to as Langer's lines, wrinkle lines, and contour lines (resulting from the interaction of two skin planes, such as the cheek-nose junction), are three distinct types of natural lines that are included within it (Fig. 3.2) [21–24].

3.3 Skin Simulants

Skin simulants are materials designed to mimic the properties of human skin for various applications in industries such as medicine, cosmetics, forensics, and product testing. The development and use of skin simulants have become essential for research, training, and testing purposes (see Fig. 3.3). The primary applications of skin simulants can be found in medical training and simulation, cosmetic and dermatological research, product testing, burn care training, prosthetics, etc. [26].

3.3.1 Medical Training and Simulation

Skin simulants are utilized in medical education to provide a realistic training environment for practicing surgical procedures such as incisions, suturing, and other dermatological techniques. Such simulants help healthcare professionals practice injection techniques and blood collection procedures. Also, skin simulants are employed in

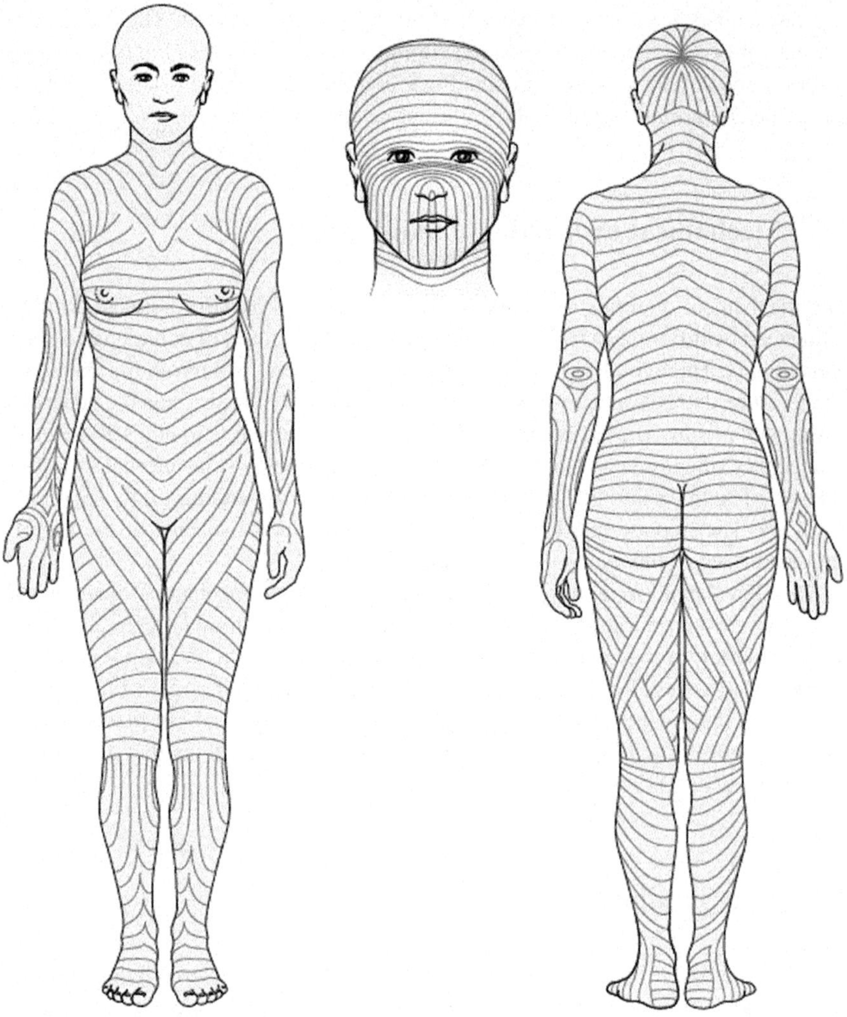

Fig. 3.2 Skin layers and natural lines [25]

wound care training to simulate various types of wounds, ulcers, and lesions. This allows healthcare professionals to practice wound assessment and management. Skin simulants are employed in training for the care of burn injuries. They help healthcare professionals practice the assessment and treatment of burns, including dressing changes and wound care [28].

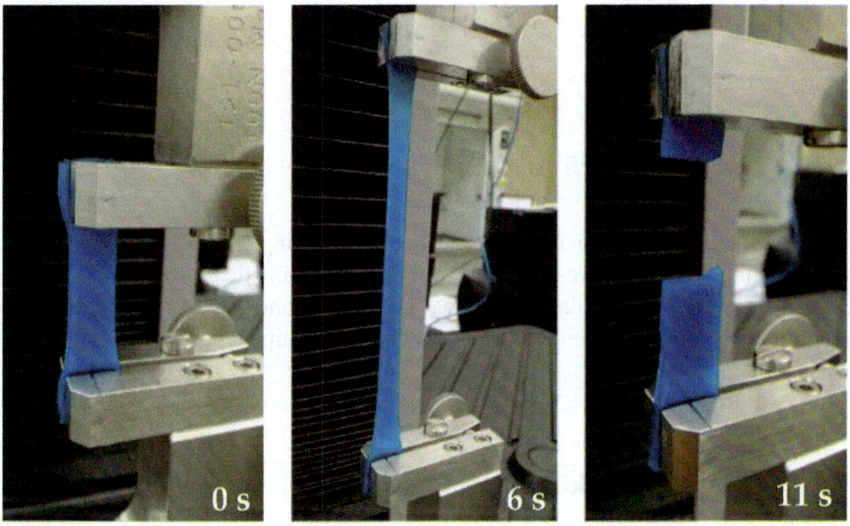

Fig. 3.3 Uniaxial test on a skin surrogate specimen captured at the beginning and during the test [27]

3.3.2 Cosmetic and Dermatological Research

Skin simulants are used in the development and testing of cosmetic products, including creams, lotions, and makeup. These simulants help assess the product's compatibility with human skin and their effects on skin texture and appearance whereas, dermatological research uses skin simulants to study skin diseases, test new treatments, and evaluate the efficacy of skincare products [29].

3.3.3 Product Testing

In industries such as textiles and apparel, skin simulants are used to test the comfort and compatibility of fabrics and materials with human skin. The development and testing of adhesive products, such as bandages and medical tapes, also involve the use of skin simulants to ensure proper adhesion and skin-friendliness [29].

3.3.4 Robotics and Prosthetics

In the development of robotic systems and prosthetics, skin simulants are used to create realistic coverings that mimic the feel and appearance of human skin. This is important for enhancing the user experience and interaction with these technologies [29].

3.3.5 *Other Research Areas*

Skin simulants play a role in forensic science for recreating and analyzing injuries and wounds in crime scene investigations. This includes ballistic testing and bloodstain pattern analysis. Skin simulants are also used in tattoo and piercing training to allow practitioners to practice their skills in a realistic environment before working on real clients [29].

3.4 Materials for Skin Simulants

Skin simulants come in various forms, including synthetic materials, gel-based formulations, polyurethane, and silicone-based materials. Creating skin simulants involves selecting materials that mimic the physical and mechanical properties of human skin. The choice of simulant depends on the specific application and the desired properties, such as texture, elasticity, and response to external stimuli [28, 29]. These simulants play a crucial role in advancing research, training, and product development while minimizing the need for direct testing on human subjects. Silicone rubber is a popular choice for skin simulants due to its flexibility, durability, and similarity to human skin in terms of elasticity and softness [30]. Platinum-cured silicone rubber is a popular choice for its flexibility, durability, and resemblance to human skin texture. Different grades of silicone rubber can be used to replicate various skin types, such as young or aged skin. The various hardness levels of silicone allow for the simulation of different skin types (e.g., soft and pliable or firm and elastic) [31, 32].

Gelatin is a natural protein derived from collagen and is often used to create simple and cost-effective skin simulants. Gelatin-based mixtures can be used to replicate the feel and texture of human skin. It can be mixed with water and other additives to achieve different textures, properties, and varying degrees of softness or firmness [27, 33, 34]. Latex is a stretchable and flexible material that can be used to simulate skin texture. Latex can be combined with other materials to create more realistic skin simulants, but it may not be suitable for individuals with latex allergies [29]. Polyurethane elastomers can be customized to mimic the mechanical properties of human skin. It is available in various forms, such as foams and elastomers, providing options for different applications. They offer good durability and can be molded into various shapes. TPEs are a class of materials that combine the characteristics of rubber and plastic. TPE materials, such as thermoplastic polyurethane (TPU), offer a balance of elasticity and flexibility, making them suitable for skin simulation. They can be engineered to resemble human skin and are often used in medical training simulations [35]. Collagen-based materials can provide a realistic feel due to their similarity to the protein found in human skin. Collagen sheets or gels can be used to create layers that simulate the different structures of the skin. These materials may include collagen hydrogels or films [36, 37]. Adding plasticizers or fillers to base materials can help achieve specific properties, such as flexibility, resilience, and texture.

When creating skin simulants, it's important to consider the specific properties to replicate, such as elasticity, color, and texture. Experimentation and customization of material formulations are often necessary to achieve the desired results for a particular application. Additionally, ethical considerations should be taken into account, especially when creating realistic skin simulants for medical or training purposes [29].

3.5 Applications of Skin Simulants

Skin simulants have various applications across different fields, including medical training, research, product development, and testing. Some key applications of skin simulants are discussed below.

3.5.1 Medical Training

Skin simulants play a vital role in medical training and simulation, providing a realistic and safe environment for healthcare professionals to practice various procedures, enhance their skills, and improve patient outcomes. Skin simulants are extensively used in surgical training programs to simulate the feel, texture, and characteristics of human skin. This allows trainees to practice incisions, suturing, and other surgical techniques in a controlled and realistic setting. Simulants can replicate different skin types and thicknesses, helping surgeons and medical students gain experience with a variety of scenarios they may encounter in real surgical procedures [26, 38].

Skin simulants can mimic various skin conditions, such as wounds, ulcers, lesions, and rashes (see Fig. 3.4). This enables medical professionals to practice diagnosing and treating different dermatological issues without putting real patients at risk. Training on skin simulants allows healthcare practitioners to practice wound assessment, cleaning, and dressing techniques. Simulated wounds can be designed to replicate different stages of healing or specific wound types. Skin simulants are employed in medical simulation scenarios to recreate emergency situations. This includes simulating trauma injuries, burns, and other critical conditions that require rapid and effective medical intervention. Using skin simulants in standardized patient encounters helps medical professionals practice communication skills related to dermatological

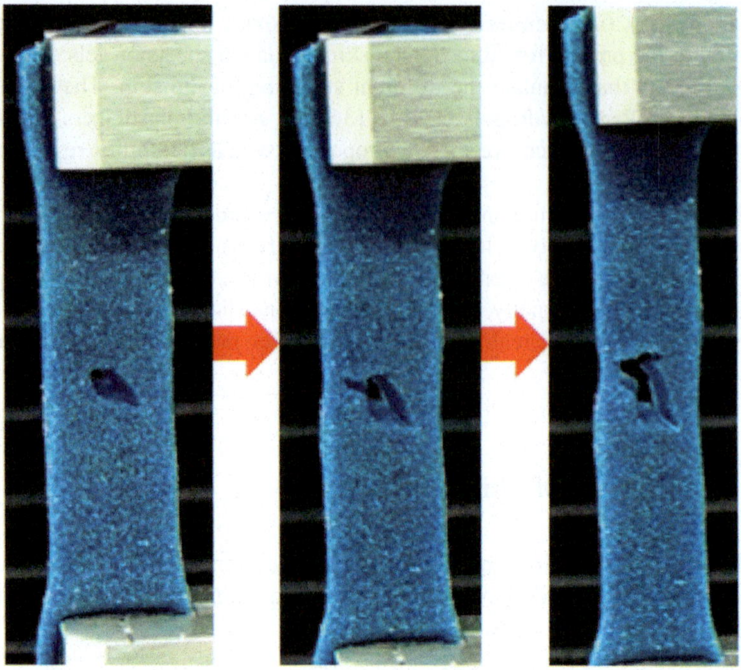

Fig. 3.4 Mechanical tensile test of a wounded skin phantom used for DIC measurements [41]

issues [32, 39, 40]. This is important for building rapport and ensuring patient understanding and compliance. High-quality skin simulants are designed to be anatomically accurate, replicating not only the visual appearance but also the underlying structures of the skin. This ensures a comprehensive learning experience for medical trainees.

In summary, skin simulants are indispensable tools in medical training and simulation, offering a risk-free and repeatable environment for healthcare professionals to refine their skills, enhance patient care, and ultimately improve medical outcomes. Their versatility makes them applicable across a wide range of medical disciplines and scenarios.

3.5.2 Testing of Cosmetic Products

Skin simulants play a crucial role in the cosmetic industry, particularly in the testing and development of cosmetic products. These simulants provide a realistic substitute for human skin, allowing researchers and cosmetic scientists to assess the safety, efficacy, and performance of various cosmetic formulations. Skin simulants are employed to assess the safety of cosmetic products by mimicking human skin reactions [24, 42, 43]. This includes testing for potential irritations, sensitivities, and allergic reactions that might occur when the cosmetic product comes into contact with the skin. Simulants are used to study how cosmetic ingredients are absorbed and penetrate the skin. This helps researchers understand the bioavailability of active ingredients, potential systemic effects, and the depth of penetration into different layers of the skin. Skin simulants allow researchers to study the permeability of cosmetic formulations. This involves assessing how well ingredients can pass through the skin barrier, influencing the effectiveness of the product in delivering active components [44, 45].

Cosmetic scientists use skin simulants to test the formulation of products, ensuring that they have the right consistency, texture, and appearance. This includes assessing creams, lotions, gels, and other formulations for their feel on the skin. Also, skin simulants are widely used to evaluate the efficacy of sunscreens in protecting the skin from UV radiation. This includes assessing factors such as Sun Protection Factor (SPF), UVA/UVB protection, and water resistance. Simulants help in testing the effectiveness of anti-aging products by mimicking the aging process of human skin. This includes evaluating the product's ability to reduce wrinkles, improve skin elasticity, and address other signs of aging. Skin simulants are utilized to test the performance of makeup products, including foundation, blush, eyeshadow, and lipstick [46–48]. This involves assessing color matching, adherence, and the overall appearance of the cosmetics on the skin. Simulants are used to evaluate the moisturizing properties of cosmetic products. This includes studying how well a moisturizer hydrates the skin, improves skin barrier function, and provides long-lasting hydration. Skin simulants offer an ethical alternative to animal testing for cosmetic products, aligning with the growing emphasis on cruelty-free practices in the cosmetics industry [29].

In summary, skin simulants are invaluable tools in the cosmetic industry, providing a reliable and ethical means to assess the safety, performance, and overall quality of cosmetic products before they are introduced to the market. Their use contributes to the development of innovative and consumer-friendly cosmetic formulations.

3.5.3 Educational Purposes

Skin simulants serve educational purposes in various fields, providing a hands-on and realistic learning experience without the need for live human subjects. Whether in medical education, first aid training, or other educational contexts, these simulants play a crucial role in enhancing understanding, skills development, and practical knowledge. Skin simulants help students study the structure and function of the skin, including the epidermis, dermis, and subcutaneous layers. Anatomically accurate models facilitate a better understanding of the skin's role in the human body [46]. Simulants allow medical and nursing students to practice various clinical skills, such as wound care, suturing, and injection techniques, in a controlled and safe environment before working with real patients. Skin simulants can be used to recreate realistic wounds and injuries for first aid and trauma training. This enables students and emergency responders to practice assessing and treating injuries in a lifelike setting [49–51]. Simulants can mimic a variety of dermatological conditions, allowing dermatology students and practitioners to practice identifying and diagnosing skin disorders without using real patients. Skin simulants provide a platform for surgical residents and students to practice making incisions, suturing wounds, and performing other surgical procedures, enhancing their technical skills and precision. Similarly, skin simulants can be used in dental education to practice oral and maxillofacial procedures, including suturing and wound care [28, 52, 53].

In summary, skin simulants contribute significantly to educational programs by providing a safe and controlled environment for learners to acquire practical skills, understand anatomical structures, and develop confidence in various medical and healthcare procedures. Their use enhances the educational experience and ensures that professionals are well-prepared for real-world scenarios.

3.5.4 Forensic Science

Skin simulants are valuable tools in forensic science for simulating human skin in various scenarios, including crime scene reconstructions, training, and research. These simulants enable forensic scientists and investigators to practice and refine their skills in a controlled environment. Skin simulants are used to recreate realistic crime scenes, including injuries, wounds, and blood spatter patterns. This helps forensic scientists understand the dynamics of violent incidents and aids in the reconstruction of events. Simulants that replicate the texture and properties of human skin are employed in bloodstain pattern analysis [9, 49, 54]. This allows investigators to study how blood behaves when it comes into contact with the skin and to accurately interpret bloodstain patterns. Simulants help forensic experts study wound ballistics by simulating gunshot wounds and other injuries. This aids in understanding the trajectory of projectiles and the extent of damage caused by different weapons. Medical examiners and forensic pathologists use skin simulants to train for autopsies and post-mortem examinations [29, 31].

Simulants provide a realistic representation of the skin's response to injuries, allowing pathologists to refine their skills. Skin simulants are used in training programs for crime scene investigators to practice the proper collection and documentation of evidence related to skin injuries and conditions. In forensic odontology, simulants may be used to simulate bite marks on the skin, allowing forensic odontologists to practice analysis and comparison techniques. Skin simulants, when accurately designed and utilized, contribute to the improvement of forensic science practices by providing a realistic and repeatable environment for training, research, and evidence analysis. Their use helps forensic professionals develop the skills necessary for accurate and reliable crime scene investigations [29].

3.5.5 Product Testing

Skin simulants are essential in product testing across various industries, ensuring the safety, efficacy, and performance of products that come into contact with the skin. These simulants provide a reliable and ethical alternative to testing on human subjects, allowing companies to assess the impact of their products without potential harm. Skin simulants are used to assess the potential irritation and sensitivity of products, including cosmetics, skincare items, and household products. This helps identify any adverse reactions that might occur when a product comes into contact with the skin [26, 27]. Simulants are employed to study how products are absorbed and penetrate the skin. This includes assessing the bioavailability of active ingredients and understanding the depth of penetration into different layers of the skin.

Simulants can be used to test the impact of products on wound healing and scar reduction. This is particularly relevant in the development of products like wound care solutions, scar creams, and post-surgery skincare. Skin simulants help in testing the effectiveness of anti-aging products by simulating the aging process of human skin. This includes evaluating a product's ability to reduce wrinkles, improve skin elasticity, and address other signs of aging. Simulants are used to evaluate the moisturizing properties of products, including lotions and creams. This includes studying how well a product hydrates the skin, improves skin barrier function, and provides long-lasting hydration. Skin simulants provide an ethical alternative to animal testing and ensure that product testing is conducted in a humane and responsible manner [55, 56].

3.5.6 Automobile Industry

Skin simulants are not commonly used in the automobile industry in the same way they are in fields like medical training or product testing. However, certain materials and technologies are employed in the automotive sector for specific purposes related to human interaction with vehicles. Automakers may use materials that simulate the feel and texture of human skin for testing purposes. This could include evaluating the comfort and durability of materials used in car interiors, such as seats, steering wheels, and armrests. Simulants may be used in the design and testing of vehicle interfaces to ensure that controls and touchpoints are comfortable for users [29]. This involves considering the tactile aspects of switches, buttons, and touchscreens to provide an intuitive and user-friendly experience.

Simulants might be used in assessing factors like pressure points, vibration resistance, and support to optimize seat design for driver and passenger comfort. In studies related to safety and injury prevention, researchers may use materials that simulate the mechanical properties of human skin to understand the potential effects of airbags, seat belts, and other safety features on occupants during a crash. While the automotive industry doesn't typically use skin simulants in the same way as some other sectors, there is a focus on materials and technologies that enhance safety, comfort, and user experience. The primary emphasis is on crash testing, comfort, and safety testing rather than replicating the intricate properties of human skin [37, 57].

References

1. Diridollou S et al (2000) In vivo model of the mechanical properties of the human skin under suction. Ski Res Technol 6(4):214–221. https://doi.org/10.1034/J.1600-0846.2000.006004214.X
2. Hochberg J, Meyer KM, Marion MD (2009) Suture choice and other methods of skin closure. Surg Clin North Am 89(3):627–641. https://doi.org/10.1016/J.SUC.2009.03.001
3. Untaroiu CD, Yue N, Shin J (2013) A finite element model of the lower limb for simulating automotive impacts. Ann Biomed Eng 41(3):513–526. https://doi.org/10.1007/S10439-012-0687-0/FIGURES/13
4. Edwards C, Marks R (1995) Evaluation of biomechanical properties of human skin. Clin Dermatol 13(4):375–380. https://doi.org/10.1016/0738-081X(95)00078-T
5. Goldsmith LA (1990) My organ is bigger than your organ. Arch Dermatol 126(3):301–302. https://doi.org/10.1001/archderm.1990.01670270033005
6. Ottenio M, Tran D, Ní Annaidh A, Gilchrist MD, Bruyère K (2015) Strain rate and anisotropy effects on the tensile failure characteristics of human skin. J Mech Behav Biomed Mater 41:241–250. https://doi.org/10.1016/j.jmbbm.2014.10.006
7. Chanda A, Graeter R, Unnikrishnan V (2015) Effect of blasts on subject-specific computational models of skin and bone sections at various locations on the human body. AIMS Mater Sci 2(4):425–447. https://doi.org/10.3934/matersci.2015.4.425
8. Groves RB, Coulman SA, Birchall JC, Evans SL (2013) An anisotropic, hyperelastic model for skin: experimental measurements, finite element modelling and identification of parameters for human and murine skin. J Mech Behav Biomed Mater 18:167–180. https://doi.org/10.1016/j.jmbbm.2012.10.021
9. Fenton LA, Horsfall I, Carr DJ (2020) Skin and skin simulants. Aust J Forensic Sci (Taylor and Francis Ltd.) 52(1):96–106. https://doi.org/10.1080/00450618.2018.1450896
10. Pailler-Mattei C, Bec S, Zahouani H (2008) In vivo measurements of the elastic mechanical properties of human skin by indentation tests. Med Eng Phys 30(5):599–606. https://doi.org/10.1016/j.medengphy.2007.06.011
11. Geerligs M, Peters GWM, Ackermans PAJ, Oomens CWJ, Baaijens FPT (2008) Linear viscoelastic behavior of subcutaneous adipose tissue. Biorheology 45(6):677–688. https://doi.org/10.3233/BIR-2008-0517
12. Miller-Young JE, Duncan NA, Baroud G (2002) Material properties of the human calcaneal fat pad in compression: experiment and theory. J Biomech 35(12):1523–1531. https://doi.org/10.1016/S0021-9290(02)00090-8
13. McGrath JA, Uitto J (2010) Anatomy and organization of human skin. In: Rook's textbook of dermatology, vol 1. Wiley-Blackwell, Oxford, pp 1–53
14. Rahmati M, Blaker JJ, Lyngstadaas SP, Mano JF, Haugen HJ (2020) Designing multigradient biomaterials for skin regeneration. Mater Today Adv 5:100051. https://doi.org/10.1016/J.MTADV.2019.100051
15. Lee Y, Hwang K (2002) Skin thickness of Korean adults. Surg Radiol Anat 24(3–4):183–189. https://doi.org/10.1007/s00276-002-0034-5
16. Silver FH, Freeman JW, Devore D (2001) Viscoelastic properties of human skin and processed dermis. Ski Res Technol 7(1):18–23. https://doi.org/10.1034/j.1600-0846.2001.007001018.x
17. Sutradhar A, Miller MJ (2013) In vivo measurement of breast skin elasticity and breast skin thickness. Ski Res Technol 19(1):e191–e199. https://doi.org/10.1111/j.1600-0846.2012.00627.x

18. Ní Annaidh A, Bruyère K, Destrade M, Gilchrist MD, Otténio M (2012) Characterization of the anisotropic mechanical properties of excised human skin. J Mech Behav Biomed Mater 5(1):139–148. https://doi.org/10.1016/j.jmbbm.2011.08.016
19. Wu JZ, Dong RG, Smutz WP, Schopper AW (2003) Nonlinear and viscoelastic characteristics of skin under compression: experiment and analysis. Biomed Mater Eng 13(4):373–385
20. Flynn C, McCormack BAO (2008) Finite element modelling of forearm skin wrinkling. Ski Res Technol 14(3):261–269. https://doi.org/10.1111/j.1600-0846.2008.00289.x
21. Langer K (1978) On the anatomy and physiology of the skin. I. The cleavability of the cutis. Br J Plast Surg 31(1):3–8. https://doi.org/10.1016/0007-1226(78)90003-6
22. Kwak M, Son D, Kim J, Han K (2014) Static Langer's line and wound contraction rates according to anatomical regions in a porcine model. Wound Repair Regen 22(5):678–682. https://doi.org/10.1111/wrr.12206
23. Cox HT (1941) The cleavage lines of the skin. Br J Surg 29(114):234–240. https://doi.org/10.1002/bjs.18002911408
24. Pissarenko A, Meyers MA (2020) The materials science of skin: analysis, characterization, and modeling. Prog Mater Sci (Elsevier Ltd.) 110:100634. https://doi.org/10.1016/j.pmatsci.2019.100634
25. Newell KA (2007) Wound closure. Essent Clin Proced 313–341. https://doi.org/10.1016/B978-1-4160-3001-0.50027-7
26. Gupta V, Singh G, Chanda A (2023) Modeling of metamaterial based incision patterns for generating high expansions in skin grafts. Clin Biomech 110:106118. https://doi.org/10.1016/J.CLINBIOMECH.2023.106118
27. Chanda A (2018) Biomechanical modeling of human skin tissue surrogates. Biomimetics 3(3):18. https://doi.org/10.3390/BIOMIMETICS3030018
28. Makode S, Singh G, Chanda A (2021) Development of novel anisotropic skin simulants. Phys Scr 96(12):125019. https://doi.org/10.1088/1402-4896/AC2EFD
29. Singh G, Chanda A (2021) Mechanical properties of whole-body soft human tissues: a review. Biomed Mater 16(6):062004. https://doi.org/10.1088/1748-605X/AC2B7A
30. Singh G, Gupta V, Chanda A (2022) Artificial skin with varying biomechanical properties. Mater Today Proc 62:3162–3166. https://doi.org/10.1016/J.MATPR.2022.03.433
31. Gupta V, Singh G, Chanda A (2022) Development and testing of skin grafts models with varying slit orientations. Mater Today Proc 62:3462–3467. https://doi.org/10.1016/J.MATPR.2022.04.282
32. Gupta V, Chanda A (2022) Biomechanics of skin grafts: effect of pattern size, spacing and orientation. Eng Res Express 4(1):015006. https://doi.org/10.1088/2631-8695/AC48CB
33. Casem DT, Dwivedi AK, Mrozek RA, Lenhart JL (2014) Compression response of a thermoplastic elastomer gel tissue surrogate over a range of strain-rates. Int J Solids Struct 51(11–12):2037–2046. https://doi.org/10.1016/J.IJSOLSTR.2013.12.028
34. Karimi A, Navidbakhsh M (2014) Material properties in unconfined compression of gelatin hydrogel for skin tissue engineering applications. Biomed Tech 59(6):479–486. https://doi.org/10.1515/BMT-2014-0028/MACHINEREADABLECITATION/RIS
35. Chhikara K, Singh G, Gupta S, Chanda A (2022) Progress of additive manufacturing in fabrication of foot orthoses for diabetic patients: a review. Ann 3D Print Med 8:100085. https://doi.org/10.1016/J.STLM.2022.100085
36. Gupta V, Chanda A (2023) Expansion potential of novel skin grafts simulants with I-shaped auxetic incisions. Biomed Eng Adv 5:100071. https://doi.org/10.1016/J.BEA.2023.100071
37. Singh G, Chanda A (2023) Development and mechanical characterization of artificial surrogates for brain tissues. Biomed Eng Adv 5:100084. https://doi.org/10.1016/J.BEA.2023.100084
38. Stevenson TC et al (2023) Descriptive analysis of forces applied by trained clinicians during 2-handed instrument-assisted soft tissue mobilization. J Athl Train 58(1):60–64. https://doi.org/10.4085/1062-6050-282-21
39. Gupta V, Chanda A (2022) Expansion potential of skin grafts with alternating slit based auxetic incisions. Forces Mech 7:100092. https://doi.org/10.1016/J.FINMEC.2022.100092

40. Singh G, Chanda A (2022) Biomechanical modeling of progressive wound healing: a computational study. Biomed Eng Adv 4:100055. https://doi.org/10.1016/J.BEA.2022.100055
41. Chanda A, Upchurch W (2018) Biomechanical modeling of wounded skin. J Compos Sci 2(4):69. https://doi.org/10.3390/JCS2040069
42. Dabrowska AK et al (2016) Materials used to simulate physical properties of human skin. Ski Res Technol 22(1):3–14. https://doi.org/10.1111/SRT.12235
43. Koudine, Barquins, Anthoine, Aubert, Lévêque (2000) Frictional properties of skin: proposal of a new approach. Int J Cosmet Sci 22(1): 11–20. https://doi.org/10.1046/J.1467-2494.2000.00006.X
44. Gefen A, Dilmoney B (2007) Mechanics of the normal woman's breast. Technol Heal Care 15(4):259–271. https://doi.org/10.3233/thc-2007-15404
45. Bhushan B, Wei G, Haddad P (2005) Friction and wear studies of human hair and skin. Wear 259(7–12):1012–1021. https://doi.org/10.1016/J.WEAR.2004.12.026
46. Guerra C, Schwartz CJ (2011) Development of a synthetic skin simulant platform for the investigation of dermal blistering mechanics. Tribol Lett 44(2):223–228. https://doi.org/10.1007/S11249-011-9840-4/TABLES/5
47. Connolly M et al (2019) Novel polylactic acid (PLA)-organoclay nanocomposite bio-packaging for the cosmetic industry; migration studies and in vitro assessment of the dermal toxicity of migration extracts. Polym Degrad Stab 168:108938. https://doi.org/10.1016/J.POLYMDEGRADSTAB.2019.108938
48. James T, Collins S, Amlôt R, Marczylo T (2019) GC–MS/MS quantification of benzyl salicylate on skin and hair: a novel chemical simulant for human decontamination studies. J Chromatogr B 1129:121818. https://doi.org/10.1016/J.JCHROMB.2019.121818
49. Annaidh AN, Bruyère K, Destrade M, Gilchrist MD, Otténio M (2012) Characterization of the anisotropic mechanical properties of excised human skin. J Mech Behav Biomed Mater 5(1):139–148. https://doi.org/10.1016/J.JMBBM.2011.08.016
50. Jussila J, Leppäniemi A, Paronen M, Kulomäki E (2005) Ballistic skin simulant. Forensic Sci Int 150(1):63–71. https://doi.org/10.1016/J.FORSCIINT.2004.06.039
51. Gupta V, Singh G, Gupta S, Chanda A (2023) Expansion potential of auxetic prosthetic skin grafts: a review. Eng Res Express 5(2):022003. https://doi.org/10.1088/2631-8695/ACCFE5
52. Cua AB, Wilhelm KP, Maibach HI (1995) Skin surface lipid and skin friction: relation to age, sex and anatomical region. Skin Pharmacol Physiol 8(5):246–251. https://doi.org/10.1159/000211354
53. Willson SA, Adam EJ, Tucker AK (1982) Patterns of breast skin thickness in normal mammograms. Clin Radiol 33(6):691–693. https://doi.org/10.1016/S0009-9260(82)80407-8
54. Gallagher AJ, Ní Anniadh A, Bruyere K, Otténio M, Xie H, Gilchrist MD (2012) Dynamic tensile properties of human skin. In: 2012 IRCOBI conference proceedings—international research council on the biomechanics of injury, pp 494–502
55. Gupta V, Singh G, Chanda A (2023) Development of hierarchical auxetic skin graft simulants with high expansion potential. Biomed Eng Adv 5:100087. https://doi.org/10.1016/J.BEA.2023.100087
56. Gupta V, Singh G, Chanda A (2023) High expansion auxetic skin graft simulants for severe burn injury mitigation. Eur Burn J 4(1):108–120. https://doi.org/10.3390/EBJ4010011
57. Abdel-Nasser YA (2013) Frontal crash simulation of vehicles against lighting columns using FEM. Alexandria Eng J 52(3):295–299. https://doi.org/10.1016/J.AEJ.2013.01.005

Chapter 4
Muscle Tissue Simulants

Muscles are flexible tissues with contractile properties and are viscoelastic, anisotropic, and nonlinear in nature. Muscle tissues are responsible for controlling body parts or organs to perform different functions [1]. Muscle tissue is one of the four basic types of tissues in the human body, along with epithelial, connective, and nervous tissues. Muscle tissue is specialized for contraction, which allows movement and force generation [2, 3]. There are three main types of muscle tissues in the human body, i.e., skeletal muscle tissue, smooth muscle tissue, and cardiac muscle tissue (Fig. 4.1).

4.1 Skeletal Muscle Tissue

Skeletal muscle tissue is a type of muscle tissue that is attached to bones by tendons and is responsible for voluntary body movements. The contraction of skeletal muscle allows for activities such as walking, running, and lifting weights. The structure of the skeletal muscle fibers are long, multinucleated cells with a striated (striped) appearance under a microscope. Skeletal muscles provide conscious, voluntary control over body movements. They contract in response to signals from the nervous system [2, 5].

4.1.1 Key Characteristics of Skeletal Muscle Tissue

Skeletal muscles are under conscious or voluntary control. Movements like walking, talking, and reaching are initiated and coordinated by signals from the central nervous system. They exhibit a striped or striated pattern when viewed under a microscope. This striation is due to the organized arrangement of the contractile protein's actin and myosin within the muscle cells [6, 7]. Skeletal muscle cells (also called muscle

A. Chanda and G. Singh, *Soft Tissue Simulants*, Biomedical Materials for Multi-functional Applications, https://doi.org/10.1007/978-981-97-3060-5_4

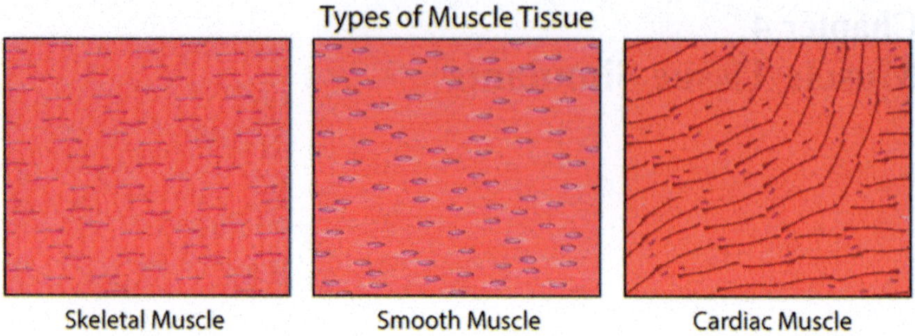

Fig. 4.1 Different types of muscles and their structures [4]

fibers) are long, cylindrical, and multinucleated. Multinucleation is a result of the fusion of multiple embryonic cells during development. Skeletal muscles are attached to bones by tendons, which are tough, fibrous connective tissues. The contraction of skeletal muscles pulls on tendons, causing movement at the joints. These muscles contract quickly but can fatigue relatively rapidly. They are well-suited for activities that require short bursts of intense effort. Muscle fibers within a skeletal muscle run parallel to each other, allowing for efficient force generation and coordinated movement [7–9]. The basic functional units of skeletal muscle are sarcomeres, repeating structural units along the length of the muscle fibers (see Fig. 4.2). Sarcomeres are the regions between two Z lines and are composed of actin and myosin filaments [6, 7].

4.1.2 *Functions of Skeletal Muscle Tissue*

Skeletal muscles are responsible for moving bones and joints, allowing for a wide range of voluntary movements such as walking, running, and lifting. These muscles work in coordination to maintain body posture and provide support against the force of gravity. The contraction of skeletal muscles generates heat, contributing to the regulation of body temperature [10, 11]. Skeletal muscles, along with the bones they are attached to, provide protection to internal organs and support the body's structure. Skeletal muscle tissue plays a role in glucose metabolism and storage, contributing to overall metabolic health [2, 12, 13].

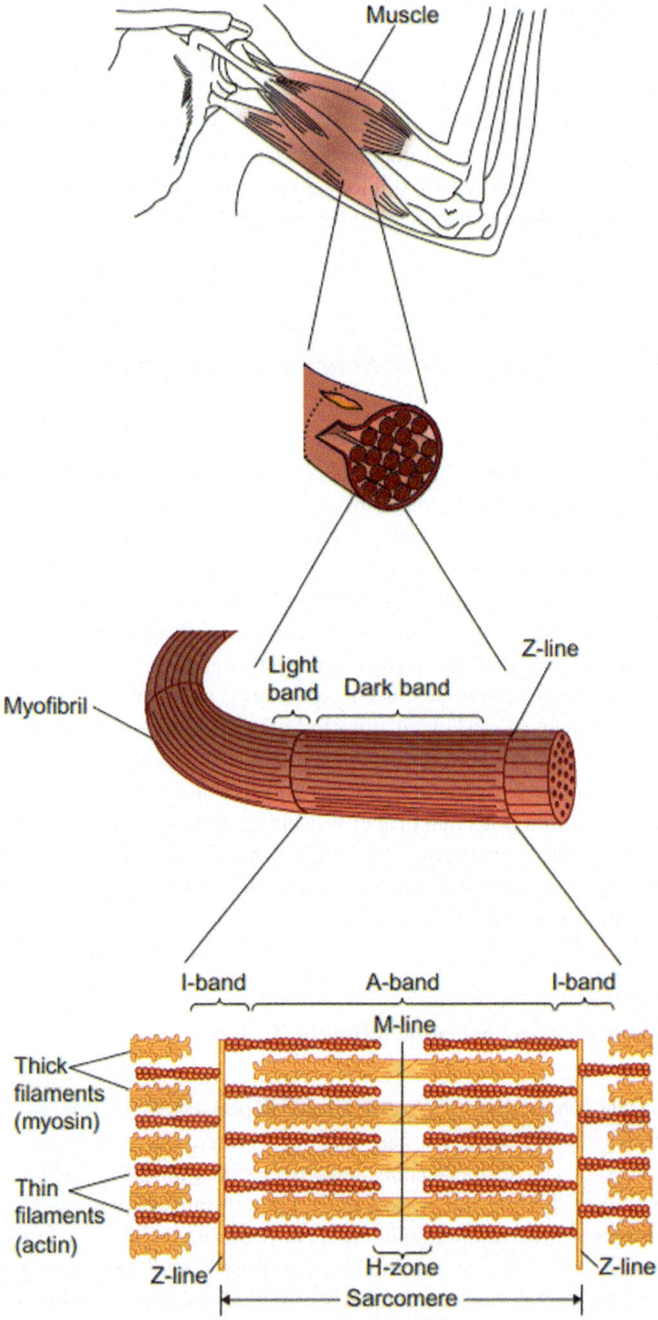

Fig. 4.2 Structure of skeletal muscle [14]

4.2 Smooth Muscle Tissue

Smooth muscle tissue is one of the three types of muscle tissues found in the human body, alongside skeletal muscle and cardiac muscle. Smooth muscle is characterized by its lack of striations (stripes) under a microscope, as well as its spindle-shaped cells. Unlike skeletal muscle, smooth muscle is involuntary, meaning it is not under conscious control, and it is responsible for various functions within internal organs [15, 16].

4.2.1 Key Characteristics of Smooth Muscle Tissue

Smooth muscle contractions are not consciously controlled; they are regulated by the autonomic nervous system. This makes smooth muscle well-suited for maintaining continuous, involuntary functions. Unlike skeletal muscle, smooth muscle lacks a striated appearance. This is due to the absence of the organized arrangement of actin and myosin filaments in the same pattern seen in skeletal muscle [17, 18]. Smooth muscle cells are elongated and spindle-shaped, with a single nucleus per cell. The lack of multinucleation distinguishes smooth muscle cells from skeletal muscle cells. Smooth muscle is found in the walls of internal organs and structures such as the digestive tract, blood vessels, respiratory passages, and reproductive organs. They contract more slowly than skeletal muscle but can sustain contractions for longer periods, making them suitable for functions such as peristalsis in the digestive tract [19, 20]. Smooth muscle cells are often organized into sheets or layers, allowing for coordinated contractions that propel substances through organs like the digestive tract. Smooth muscle is responsive to stretch, which allows for the accommodation of varying volumes of internal organs, such as the expansion of the bladder or the stretching of blood vessels [21, 22].

4.2.2 Functions of Smooth Muscle Tissue

Smooth muscle contractions play a crucial role in peristalsis, the wavelike movement that propels substances through organs like the digestive tract and the urinary system. The blood vessel walls of smooth muscle help regulate blood flow and blood pressure by constricting or dilating the vessels [21]. Smooth muscle in the walls of bronchi and bronchioles is involved in regulating airflow during respiration. Smooth muscle is found in the walls of the uterus and helps facilitate childbirth. It is also present in the reproductive organs, such as the walls of the fallopian tubes. Smooth muscle in the walls of the bladder contracts to expel urine during urination [23, 24]. Smooth muscle in the walls of the digestive organs, such as the stomach and intestines, helps mix and move food through the gastrointestinal tract [25].

4.3 Cardiac Muscle Tissue

Cardiac muscle tissue is a specialized type of muscle tissue found in the heart. It is unique in its structure and function, combining features of both skeletal and smooth muscle tissues. The contraction of cardiac muscle is involuntary, and it is responsible for pumping blood throughout the circulatory system [26].

4.3.1 Key Characteristics of Smooth Muscle Tissue

Similar to smooth muscle, cardiac muscle is under involuntary control. Contractions are regulated by the autonomic nervous system, specifically the cardiac conduction system. Cardiac muscle exhibits a striated appearance, similar to skeletal muscle. This striation is due to the organized arrangement of actin and myosin filaments within the muscle cells [26–28]. Cardiac muscle cells are branched and interconnected. This branching allows for coordinated contractions and the propagation of electrical signals between cells. Each cardiac muscle cell typically has a single nucleus, unlike skeletal muscle cells, which are multinucleated. Cardiac muscle cells are connected by specialized junctions called intercalated disks. Intercalated disks contain gap junctions that allow for the rapid transmission of electrical impulses between cells, ensuring synchronous contraction of the entire heart [29–32]. Cardiac muscle contracts rhythmically and continuously throughout a person's life. The heart beats to pump blood, and the cardiac muscle cells work in a coordinated manner to maintain this function.

4.3.2 Functions of Cardiac Muscle Tissue

The primary function of cardiac muscle is to contract and pump blood throughout the circulatory system. The heart's rhythmic contractions maintain blood circulation, delivering oxygen and nutrients to the body's tissues and organs. By regulating the force and frequency of contractions, cardiac muscle helps maintain blood pressure within the arteries [33, 34]. The electrical impulses that initiate and coordinate cardiac muscle contractions are generated by the heart's conduction system. This system ensures a synchronized and efficient pumping action. The cardiac muscle is involved in the opening and closing of heart valves. Valves ensure the unidirectional flow of blood through the heart chambers [26, 30, 35]. Cardiac muscle can adapt to changes in workload by adjusting its contractile strength and heart rate. This allows the heart to meet the varying demands of the body [29].

4.4 Muscle Tissue Simulants

Muscle tissue simulants are materials designed to mimic the mechanical and physical properties of natural muscle tissue. These simulants serve various purposes in research, medical training, and product development. Creating accurate muscle tissue simulants is important for replicating realistic conditions, allowing for effective testing, training, and experimentation [21, 36].

4.4.1 Considerations in Muscle Tissue Simulants

Simulants should replicate the elasticity, strength, and contractile properties of natural muscle tissue. This involves mimicking the behavior of muscle fibers during stretching, contracting, and relaxing. Realistic tactile feedback is crucial for training and testing applications. Simulants should provide a feel similar to that of natural muscle tissue to enhance the realism of hands-on experiences. Natural muscles exhibit viscoelastic behavior, meaning they have both elastic and viscous properties. Simulants need to replicate this behavior to accurately represent the dynamic response of muscle tissue [21, 37–39]. Muscle tissue simulants may need to respond to temperature changes similarl to real muscle tissue. This is particularly important in medical training scenarios where the temperature of tissues can affect procedures and outcomes. If the simulant is intended for medical or biological applications, it should be biocompatible to ensure safety when interacting with living tissues. Simulants may need to be customizable to represent different types of muscles, such as skeletal, smooth, and cardiac muscles, each with distinct properties [2, 13, 22, 24, 29, 34].

The development of muscle tissue simulants continues to evolve with advancements in materials science and bioengineering. These simulants play a crucial role in medical training, biomechanical research, and the testing of medical devices and interventions. The choice of a specific muscle tissue simulant depends on the intended application and the properties required for the simulation.

4.5 Materials for Muscle Tissue Simulants

Creating muscle tissue simulants involves selecting materials that closely mimic the mechanical and physical properties of natural muscle tissue. The ideal material should exhibit elasticity, viscoelasticity, and response to strain rates similar to actual muscle tissue. Various synthetic and natural materials, or combinations thereof, are used to formulate muscle tissue simulants [40, 41].

- ***Silicone Elastomers***: Silicone elastomers, such as polydimethylsiloxane (PDMS), are widely used in the creation of muscle tissue simulants. They can be formulated to exhibit elastic behavior, making them suitable for mimicking the stretching and contracting properties of muscle tissue [36].
- ***Thermoplastic Elastomers (TPEs)***: TPEs, including materials like thermoplastic polyurethane (TPU), offer a balance of flexibility and elasticity. TPEs can be engineered to resemble the mechanical properties of muscle tissue and are used in various applications, including medical simulations [42].
- ***Hydrogels***: Hydrogels are water-absorbing materials with a high water content, making them suitable for simulating the hydration properties of muscle tissue. They can be tuned to exhibit viscoelastic behavior and are commonly used in tissue engineering [43].
- ***Polyurethane (PU) Foams***: Polyurethane foams can be engineered to mimic the softness and compressibility of natural muscle tissue. These foams are often used in prosthetics and medical training models [44].
- ***Natural Rubber Latex***: Natural rubber latex can be used to create stretchable materials that resemble the elasticity of muscle tissue. It provides a good balance between flexibility and resilience [45, 46].
- ***Elastomeric Polymers***: Various elastomeric polymers, such as styrene–butadiene–styrene (SBS) or styrene-ethylene-butylene-styrene (SEBS), can be tailored to exhibit the mechanical properties required for muscle tissue simulants [47].
- ***Gelatin-Based Models***: Gelatin, a protein derived from collagen, is sometimes used to create soft tissue simulants. It can be modified to exhibit certain mechanical properties and is often employed in educational settings for hands-on training [48, 49].
- ***Synthetic Polymeric Blends***: Researchers may combine different synthetic polymers to create blends that offer a more comprehensive representation of muscle tissue behavior. This approach allows for customization of material properties [50].
- ***3D-Printed Materials***: Advances in 3D printing technology enable the creation of muscle tissue simulants using various printable materials. Bioinks, hydrogels, and other printable materials can be used to replicate the complex structure of muscle tissue [51].

The choice of material depends on the specific requirements of the application, the desired properties of the muscle tissue simulant, and the intended use (such as medical training, biomechanical research, and prosthetics development). Experimentation and validation against real muscle tissue properties are essential to ensure that the simulants accurately represent the behavior of natural muscles. Additionally, ongoing research in materials science and tissue engineering may lead to the development of more sophisticated muscle tissue simulants in the future.

4.6 Fabrication of Muscle Tissue Simulants

Across all the materials for fabricating muscle tissue simulants, silicone elastomers and hydrogels are the most widely used materials. Creating muscle tissue simulants using silicone elastomers and hydrogels involves a careful formulation that allows for the replication of the mechanical and physical properties of natural muscle tissue. Silicone elastomers are well-suited for this purpose due to their flexibility, durability, and ability to mimic the elasticity of muscle tissue. Hydrogels are water-absorbing materials with high water content, making them suitable for simulating the hydration properties of soft tissues [21, 36].

4.6.1 Silicone Elastomers

The first step is to measure the silicone elastomer base and curing agent and, if using a specific formulation that requires a catalyst, add it to the solution and thoroughly mix the components. We can also add pigments or dyes to the silicone mixture and mix until the color is uniform to have a specific color for the muscle tissue simulant. To minimize the presence of air bubbles in the final product, the mixed silicone can be degassed. Place the mixture in a vacuum chamber to remove trapped air. This step is optional but can enhance the quality of the simulants [37]. Then, use a mold designed to shape the muscle tissue simulant, and carefully pour the mixed silicone into the prepared mold ensuring that the silicone fills the mold evenly. After pouring the silicone solution, allow the silicone to cure or set according to the recommended curing time provided by the manufacturer. The curing process may involve room temperature curing or the use of heat, depending on the specific silicone formulation. Once the silicone has fully cured, carefully demold the muscle tissue simulant. Trim any excess material or refine the simulated muscle tissue as needed to achieve the desired shape and appearance [38, 39, 52, 53]. It's important to note that the specific details of the process may vary based on the type and formulation of the silicone elastomer used. Additionally, iterative testing and refinement may be necessary to achieve the desired properties in the muscle tissue simulants.

4.6.2 Hydrogels

The first step is to choose a hydrogel precursor suitable for muscle tissue simulation. Common hydrogel materials include polyacrylamide, polyethylene glycol, and agarose. Mix the hydrogel precursor with water or a buffer solution and stir the solution thoroughly to ensure homogeneity. The next step is to choose a mold that represents the desired shape of the simulated muscle tissue [21]. Apply a release agent to the mold to prevent the hydrogel from sticking. Carefully pour the prepared hydrogel solution into the mold, ensuring that it fills the mold uniformly. Allow the hydrogel to cure or set according to the recommended time and conditions provided by the manufacturer. The curing process may involve room temperature curing or temperature control, depending on the specific hydrogel formulation [54, 55]. Once the hydrogel has fully set or cured, carefully demold the simulated muscle tissue. Take care to avoid damaging the hydrogel structure. Finally, trim any excess material or refine the simulated muscle tissue as needed to achieve the desired shape and appearance.

It should be noted that if the chosen hydrogel precursor requires crosslinking, add the appropriate crosslinking agent to the solution, which helps stabilize the hydrogel structure. Some hydrogel formulations may require pH adjustment. Use pH adjusters to ensure that the hydrogel solution is within the desired pH range.

4.7 Effect of Strain Rates on the Muscle Tissue Simulants

The effect of strain rates on muscle tissue simulants is a critical consideration in biomechanics, materials science, and medical simulations. Strain rate refers to the rate at which a material deforms or experiences deformation. In the context of muscle tissue simulants, understanding how these materials respond to different strain rates is essential for accurately representing physiological conditions, optimizing medical training models, and designing effective biomechanical experiments [21]. There are several aspects to be considered and some of these are discussed below.

Dynamic Behavior: Muscle tissues in the human body experience a wide range of strain rates during various activities. For example, the strain rate during a slow stretch is significantly different from the rate during a sudden impact. Simulants should be designed to capture this dynamic behavior accurately [21].

Viscoelastic Properties: Muscles exhibit viscoelastic behavior, meaning their response is influenced by both elastic (instantaneous deformation) and viscous (time-dependent deformation) components. Simulants should be formulated to replicate these properties across different strain rates [21].

Material Properties: The mechanical properties of muscle tissue simulants, such as stiffness, modulus of elasticity, and ultimate tensile strength, may vary with strain rate. Understanding how these properties change can help in selecting or designing simulants that closely mimic real muscle tissue behavior [21].

Impact on Training Simulations: In medical training simulations, the accurate representation of muscle behavior under various conditions is crucial. Simulants should provide realistic feedback to practitioners, considering the strain rates associated with different medical procedures or traumatic events [21].

Biomechanical Research: Researchers studying biomechanics and injury mechanisms need to consider the impact of strain rates on muscle tissue simulants when conducting experiments. This is particularly relevant in studies involving impact, collisions, or rapid movements [21].

Validation and Calibration: Muscle tissue simulants often undergo validation and calibration processes to ensure their accuracy. These processes may involve testing the simulants at different strain rates to match their behavior with real muscle tissues [21].

Material Testing Equipment: The equipment used to test muscle tissue simulants, such as mechanical testing machines or impact testing devices, should be capable of applying controlled strain rates to accurately simulate real-world conditions [21].

Applications in Prosthetics and Orthotics: In the development and testing of prosthetics and orthotic devices, understanding how muscle tissue simulants respond to different strain rates is crucial for designing products that can adapt to varying movement speeds and intensities [21].

Understanding the strain rate-dependent behavior of muscle tissue simulants requires a combination of experimental testing and computational modeling. Material scientists and biomechanics researchers often use techniques such as tensile testing, compression testing, and impact testing to assess the response of simulants across different strain rates. Accurate modeling of strain rate-dependent behavior contributes to the development of more realistic muscle tissue simulants for a wide range of applications in medicine, biomechanics, and materials engineering.

References

1. Payne T, Mitchell S, Bibb R, Waters M (2015) Development of novel synthetic muscle tissues for sports impact surrogates. J Mech Behav Biomed Mater 41:357–374. https://doi.org/10.1016/j.jmbbm.2014.08.011
2. Bach AD, Beier JP, Stern-Staeter J, Horch RE (2004) Skeletal muscle tissue engineering. J Cell Mol Med 8:413–422. https://doi.org/10.1111/J.1582-4934.2004.TB00466.X
3. Hoppeler H, Vogt M (2001) Muscle tissue adaptations to hypoxia. J Exp Biol 204:3133–3139. https://doi.org/10.1242/JEB.204.18.3133
4. Smooth, skeletal, and cardiac muscles, n.d.
5. Levenberg S, Rouwkema J, Macdonald M, Garfein ES, Kohane DS, Darland DC, et al (2005) Engineering vascularized skeletal muscle tissue. Nat Biotechnol 237(23):879–884. https://doi.org/10.1038/nbt1109.
6. Frontera WR, Ochala J (2015) Skeletal muscle: a brief review of structure and function. Behav Genet 45:183–195. https://doi.org/10.1007/S00223-014-9915-Y/FIGURES/5
7. Valentine BA (2017) Skeletal muscle. Pathol Basis Vet Dis Expert Consult:908–953.e1. https://doi.org/10.1016/B978-0-323-35775-3.00015-1.
8. Cardinet GH (1997) Skeletal muscle function. Clin Biochem Domest Anim:407–440. https://doi.org/10.1016/B978-012396305-5/50017-8.
9. Hopkins PM (2006) Skeletal muscle physiology. Contin Educ Anaesth Crit Care Pain 6:1–6. https://doi.org/10.1093/BJACEACCP/MKI062
10. Ernster L, Nordenbrand K (1967) Skeletal muscle mitochondria. Methods Enzymol 10:86–94. https://doi.org/10.1016/0076-6879(67)10017-7
11. Buckingham M (2001) Skeletal muscle formation in vertebrates. Curr Opin Genet Dev 11:440–448. https://doi.org/10.1016/S0959-437X(00)00215-X
12. Mukund K, Subramaniam S (2020) Skeletal muscle: a review of molecular structure and function, in health and disease. Wiley Interdiscip Rev Syst Biol Med 12:e1462. https://doi.org/10.1002/WSBM.1462

13. Sheffield-Moore M, Urban RJ (2004) An overview of the endocrinology of skeletal muscle. Trends Endocrinol Metab 15:110–115. https://doi.org/10.1016/j.tem.2004.02.009
14. Hill JA, Olson EN (2012) An introduction to muscle. Muscle, vol 1. Elsevier Inc., pp 3–9 https://doi.org/10.1016/B978-0-12-381510-1.00001-6.
15. Webb RC (2003) Smooth muscle contraction and relaxation. Am J Physiol - Adv Physiol Educ 27:201–206. https://doi.org/10.1152/ADVAN.00025.2003/ASSET/IMAGES/LARGE/U10430173002.JPEG
16. Hafen BB, Burns B (2018) Physiology, smooth muscle. StatPearls
17. Doran AC, Meller N, McNamara CA (2008) Role of smooth muscle cells in the initiation and early progression of atherosclerosis. Arterioscler Thromb Vasc Biol 28:812–819. https://doi.org/10.1161/ATVBAHA.107.159327
18. Berridge MJ, Berridge MJ (2008) Smooth muscle cell calcium activation mechanisms. J Physiol 586:5047–5061. https://doi.org/10.1113/JPHYSIOL.2008.160440
19. Herrera AM, McParland BE, Bienkowska A, Tait R, Paré PD, Seow CY (2005) 'Sarcomeres' of smooth muscle: functional characteristics and ultrastructural evidence. J Cell Sci 118:2381–2392. https://doi.org/10.1242/JCS.02368
20. Lacolley P, Regnault V, Nicoletti A, Li Z, Michel JB (2012) The vascular smooth muscle cell in arterial pathology: a cell that can take on multiple roles. Cardiovasc Res 95:194–204. https://doi.org/10.1093/CVR/CVS135
21. Singh G, Chanda A (2021) Mechanical properties of whole-body soft human tissues: a review. Biomed Mater 16:062004. https://doi.org/10.1088/1748-605X/AC2B7A
22. Sanders KM, Kito Y, Hwang SJ, Ward SM (2016) Regulation of gastrointestinal smooth muscle function by interstitial cells. Physiology 31:316–326. https://doi.org/10.1152/PHYSIOL.00006.2016/ASSET/IMAGES/LARGE/PHY0041603390002.JPEG
23. Griendling KK, Gerthoffer WT (2007) Mechanisms of vascular smooth muscle cell migration. Circ Res 100:607–621. https://doi.org/10.1161/01.RES.0000258492.96097.47
24. Liu Z, Khalil RA (2018) Evolving mechanisms of vascular smooth muscle contraction highlight key targets in vascular disease. Biochem Pharmacol 153:91–122. https://doi.org/10.1016/J.BCP.2018.02.012
25. Montero D, Walther G, Pérez-Martin A, Vicente-Salar N, Roche E, Vinet A (2013) Vascular smooth muscle function in type 2 diabetes mellitus: a systematic review and meta-analysis. Diabetologia 56:2122–2133. https://doi.org/10.1007/S00125-013-2974-1/FIGURES/4
26. Hunter PJ, McCulloch AD, Ter Keurs HEDJ (1998) Modelling the mechanical properties of cardiac muscle. Prog Biophys Mol Biol 69:289–331. https://doi.org/10.1016/S0079-6107(98)00013-3
27. Pinnell J, Frca C, Turner S, Mb B, Howell S, Frca M (2007) Cardiac muscle physiology. Contin Educ Anaesth Crit Care Pain 7:85–88. https://doi.org/10.1093/BJACEACCP/MKM013
28. Allen DG, Kentish JC (1985) The cellular basis of the length-tension relation in cardiac muscle. J Mol Cell Cardiol 17:821–840. https://doi.org/10.1016/S0022-2828(85)80097-3
29. Stevens KR, Kreutziger KL, Dupras SK, Korte FS, Regnier M, Muskheli V et al (2009) Physiological function and transplantation of scaffold-free and vascularized human cardiac muscle tissue. Proc Natl Acad Sci U S A 106:16568–16573. https://doi.org/10.1073/PNAS.0908381106/SUPPL_FILE/0908381106SI.PDF
30. Reinecke H, MacDonald GH, Hauschka SD, Murry CE (2000) Electromechanical coupling between skeletal and cardiac muscle: Implications for infarct repair. J Cell Biol 149:731–740. https://doi.org/10.1083/JCB.149.3.731/VIDEO-1
31. Garibaldi S, Brunelli C, Bavastrello V, Ghigliotti G, Nicolini C (2005) Carbon nanotube biocompatibility with cardiac muscle cells. Nanotechnology 17:391. https://doi.org/10.1088/0957-4484/17/2/008
32. Mathur AB, Collinsworth AM, Reichert WM, Kraus WE, Truskey GA (2001) Endothelial, cardiac muscle and skeletal muscle exhibit different viscous and elastic properties as determined by atomic force microscopy. J Biomech 34:1545–1553. https://doi.org/10.1016/S0021-9290(01)00149-X

33. He JQ, Ma Y, Lee Y, Thomson JA, Kamp TJ (2003) Human embryonic stem cells develop into multiple types of cardiac myocytes. Circ Res 93:32–39. https://doi.org/10.1161/01.RES.0000080317.92718.99
34. Engelmayr GC, Cheng M, Bettinger CJ, Borenstein JT, Langer R, Freed LE (2008) Accordion-like honeycombs for tissue engineering of cardiac anisotropy. Nat Mater 7(12):1003–1010. https://doi.org/10.1038/nmat2316
35. Zimmermann WH, Didié M, Döker S, Melnychenko I, Naito H, Rogge C et al (2006) Heart muscle engineering: an update on cardiac muscle replacement therapy. Cardiovasc Res 71:419–429. https://doi.org/10.1016/J.CARDIORES.2006.03.023/2/71-3-419-FIG4.GIF
36. Makode S, Singh G, Chanda A (2021) Development of novel anisotropic skin simulants. Phys Scr 96:125019. https://doi.org/10.1088/1402-4896/AC2EFD
37. Chanda A, Unnikrishnan V, Lackey K, Robbins J (2020) Biofidelic conductive soft tissue surrogates. Int J Polym Mater Polym Biomater 69:127–135. https://doi.org/10.1080/00914037.2018.1552856
38. Gupta V, Singh G, Chanda A (2022) Development and testing of skin grafts models with varying slit orientations. Mater Today Proc 62:3462–3467. https://doi.org/10.1016/J.MATPR.2022.04.282
39. Gupta V, Chanda A (2022) Biomechanics of skin grafts: effect of pattern size, spacing and orientation. Eng Res Express 4:015006. https://doi.org/10.1088/2631-8695/AC48CB
40. Gupta V, Chanda A (2022) Expansion potential of skin grafts with alternating slit based auxetic incisions. Forces Mech 7:100092. https://doi.org/10.1016/J.FINMEC.2022.100092
41. Singh G, Chanda A (2022) Biomechanical modeling of progressive wound healing: a computational study. Biomed Eng Adv 4:100055. https://doi.org/10.1016/J.BEA.2022.100055
42. Gupta V, Singh G, Chanda A (2023) Modeling of metamaterial based incision patterns for generating high expansions in skin grafts. Clin Biomech 110:106118. https://doi.org/10.1016/J.CLINBIOMECH.2023.106118
43. Chanda A, Singh G (2023) Tissues in functional organs—low stiffness. Mater Horizons From Nat to Nanomater:33–48. https://doi.org/10.1007/978-981-99-2225-3_4/COVER.
44. Chanda A, Unnikrishnan V, Flynn Z, Lackey K (2017) Experimental study on tissue phantoms to understand the effect of injury and suturing on human skin mechanical properties. Proc Inst Mech Eng Part H J Eng Med 231:80–91. https://doi.org/10.1177/0954411916679438/ASSET/IMAGES/LARGE/10.1177_0954411916679438-FIG16.JPEG
45. Chanda A (2018) Biomechanical modeling of human skin tissue surrogates. Biomimetics 3:18. https://doi.org/10.3390/BIOMIMETICS3030018
46. Singh G, Chanda A (2023) Development and biomechanical testing of artificial surrogates for vaginal tissue. Adv Mater Process Technol. https://doi.org/10.1080/2374068X.2023.2198837
47. Gupta V, Singh G, Chanda A (2023) Development of novel hierarchical designs for skin graft simulants with high expansion potential. Biomed Phys Eng Express 9:035024. https://doi.org/10.1088/2057-1976/ACC661
48. Gupta V, Singh G, Chanda A (2023) High expansion auxetic skin graft simulants for severe burn injury mitigation. Eur Burn J 4:108–20. https://doi.org/10.3390/EBJ4010011
49. Gupta V, Singla R, Singh G, Chanda A (2023) Development of soft composite based anisotropic synthetic skin for biomechanical testing. Fibers 11:55. https://doi.org/10.3390/FIB11060055.
50. Singh G, Chanda A (2023) Development and mechanical characterization of artificial surrogates for brain tissues. Biomed Eng Adv 5:100084. https://doi.org/10.1016/J.BEA.2023.100084
51. Chhikara K, Singh G, Gupta S, Chanda A (2022) Progress of additive manufacturing in fabrication of foot orthoses for diabetic patients: a review. Ann 3D Print Med 8:100085. https://doi.org/10.1016/J.STLM.2022.100085
52. Singh G, Gupta V, Chanda A (2022) Artificial skin with varying biomechanical properties. Mater Today Proc 62:3162–3166. https://doi.org/10.1016/J.MATPR.2022.03.433
53. Chanda A, McClain S (2019) Mechanical modeling of healthy and diseased calcaneal fat pad surrogates. Biomimetics 4:1. https://doi.org/10.3390/BIOMIMETICS4010001
54. Singh G, Chanda A (2023) Biofidelic tongue and tonsils tissue surrogates. Mater Horizons From Nat to Nanomater: Part F1471:159–70. https://doi.org/10.1007/978-981-99-5064-5_10/COVER

55. Singh G, Chanda A (2023) Development and biomechanical testing of human stomach tissue surrogates. Mater Horizons From Nat to Nanomater: Part F1471:113–25. https://doi.org/10.1007/978-981-99-5064-5_7/COVER

Chapter 5
Brain Tissue Simulants

The human brain is an elastic and complex organ situated in the cranial cavity and shielded by the skull bones. Its functioning has been shown to be non-linear and viscoelastic in nature [1, 2]. The brain, weighing around three pounds and comprising nearly 100 billion neurons, serves as the primary neurological system of the human body. The brain is responsible for a wide range of cognitive functions, including perception, memory, language, decision-making, and motor control. Brain tissue is composed of various cell types, including neurons and glial cells, which work together to enable the complex functions of the brain. At the microscopic level, brain tissue is organized into various layers, regions, and nuclei. Neurons form intricate networks and circuits, and their connections, or synapses, allow for the transmission of signals [3–5]. The meninges are a series of membranous layers that wrap and protect the brain, facilitating its immersion in cerebrospinal fluid [6]. The cerebral cortex may be physically differentiated from the rest of the brain due to the presence of white matter covering the outermost layer of gray matter (see Fig. 5.1). The gray matter, composed mainly of cell bodies and dendrites, contrasts with the white matter, which consists of myelinated axons that facilitate communication between different brain regions [7, 8].

The brain is divided into different regions, each associated with specific functions. These include the cerebrum, cerebellum, and brainstem. Cerebrum, the largest part of the brain, is responsible for higher cognitive functions such as thinking, perception, and voluntary muscle movement [9–11]. Located at the back of the brain, the cerebellum is involved in coordination, balance, and fine motor control. The brainstem connects the brain to the spinal cord and includes the medulla oblongata, pons, and midbrain. It regulates essential functions such as breathing, heart rate, and basic bodily functions [12]. Understanding the structure and function of brain tissue is essential for advancing neuroscience, medicine, and our knowledge of the complexities of the human mind. Researchers use various techniques, including neuroimaging, histology, and molecular biology, to study brain tissue and gain insights into neurological disorders and cognitive processes.

A. Chanda and G. Singh, *Soft Tissue Simulants*, Biomedical Materials for Multi-functional Applications, https://doi.org/10.1007/978-981-97-3060-5_5

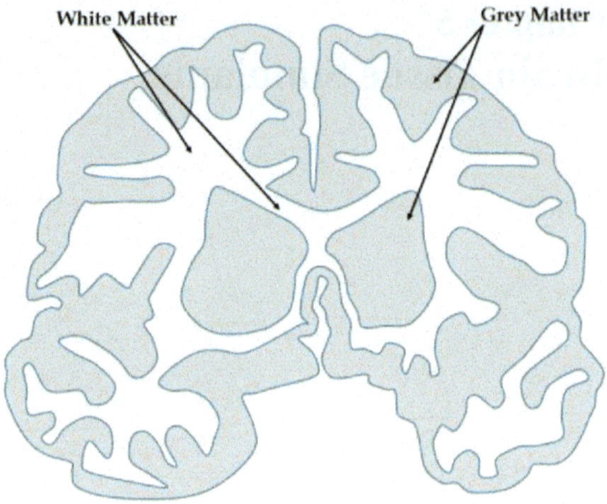

Fig. 5.1 Cross section of a brain showing gray and white matter [8]

5.1 Traditional Testing of Brain Tissues

Compression studies are often used to investigate the mechanical properties of brain tissues [13]. This is due to the complexity of the experimental testing, which includes holding the sample. When attempting to define the mechanical characteristics of brain tissues, tensile tests are rarely used [10, 14, 15]. There have only been a few studies that have documented the tensile properties of brain tissue before. Even though failure limits appear in the range of 20–60% strain [16], there is a lack of widespread comprehension of them. White matter and gray matter in the brain tissue were tested by Huang et al. [17] at frequencies of 40, 50, and 60 Hz, and the researchers discovered that the elastic modulus of white matter and gray matter was considerably different from one another. Specifically, the modulus values for white matter were found to be 3.36 ± 0.111 kPa, 3.78 ± 0.15 kPa, and 3.85 ± 0.12 kPa. On the other hand, the modulus values for gray matter were found to be 2.24 ± 0.14 kPa, 2.82 ± 0.16 kPa, and 3.33 ± 0.14 kPa, respectively. However, Budday et al. [18] reported the average value for the elastic modulus of brain tissue. In comparison to the elastic modulus of gray matter, which was determined to be 1.389 ± 0.289 kPa, the estimated elastic modulus of white matter was found to be 1.895 ± 0.592 kPa.

Another research conducted an experiment in which the brain tissue was subjected to compression loading conditions. The results of this experiment revealed that the compressive modulus of human brain tissue was 10.34 kPa [19]. The nominal compressive stress was determined by using different strain rates of 30, 60, and 90/s. The results revealed that the nominal compressive stress was 8.83 ± 1.944 kPa, 12.8 ± 3.10 kPa, and 16.0 ± 1.41 kPa, respectively, with regard to the strain rates [11]. White matter in the brain tissue showed more anisotropy than gray matter, and when exposed to higher strain rates, it demonstrated 1.3 times more stiffness [3, 20]. This was in contrast to gray matter, which exhibited lower rigidity.

It was found in the literature that brain tissue is usually tested under compression loading for mechanical properties. Compression loading is relevant to real-world scenarios where the brain may experience compression forces. This is particularly important in studies related to traumatic brain injuries (TBIs) caused by impact or external forces [18, 21]. TBIs often involve compression forces, such as those experienced during a fall, collision, or impact. Studying brain tissue under compression helps researchers understand how the tissue responds to these forces and the resulting damage [9, 22].

5.2 Brain Tissue Simulants

Creating brain tissue simulants involves formulating materials that mimic the mechanical and physical properties of natural brain tissue. These simulants find applications in various fields such as medical training, neuroscience research, and the development of medical devices. Accurate replication of brain tissue properties is crucial for realistic simulations [23]. Here are some considerations and materials commonly used for making brain tissue simulants.

5.2.1 Considerations

Simulants should mimic the anatomical and physiological features of real brain tissue as closely as possible. Mimicking the mechanical properties of brain tissue, such as elasticity and stiffness, is important for accurate representation. These properties can affect how the tissue responds to external forces and interactions. Brain tissue exhibits anisotropic properties, meaning its mechanical properties vary with direction. Simulants may attempt to replicate this anisotropy for more accurate representation [22, 24]. Since the brain is an electrically active organ, simulating its electrical conductivity is essential. This involves replicating the electrical properties of neurons and synapses. The chemical composition of the brain tissue, including neurotransmitters, ions, and other molecular components, should be considered. This could be helpful and crucial for studies involving drug interactions or biochemical processes [25, 26]. Incorporating a vascular network is important to simulate blood flow and nutrient exchange, which plays a vital role in the brain's function and health. If the simulant is intended for use in biological experiments, it should be biocompatible to ensure that it does not adversely affect living tissues. Simulants should be compatible with imaging techniques such as MRI, CT scans, and microscopy to facilitate research and analysis [27]. Stability over time is essential for experiments or studies that require long-term observation or testing. The ability to adjust the properties of the simulant based on the specific requirements of the experiment or study is important. Ethical considerations are crucial, especially if the simulants are used in research that involves testing new technologies or interventions. Depending on the application, compliance with relevant regulations and ethical guidelines is essential, especially in medical research. Considerations for cost and availability of materials used in the simulant, as well as the overall cost of producing and maintaining the simulant [28–30].

5.2.2 Materials Used for Brain Tissue Simulants

Gelatin, combined with other materials, can be used to create brain tissue simulants with tunable mechanical properties. These simulants may be suitable for educational settings and basic training. Silicone elastomers, like polydimethylsiloxane (PDMS), are versatile materials used in the development of brain tissue simulants. They can be customized to mimic the softness and compliance of brain tissue [31–33]. Hydrogels, made from polymers such as agarose or polyacrylamide, are water-absorbing materials that can replicate the water content of brain tissue. They are often used for neuroscience research and medical training. Polyurethane foams can be engineered to replicate the compressibility and softness of brain tissue [34]. They can be engineered to have similar mechanical properties to brain tissue and are often used in crash test simulations and biomechanics research. Polyvinyl Alcohol (PVA) hydrogels can be tailored to exhibit properties similar to brain tissue. They are biocompatible and can be used in medical simulations. They are also known for their stability over time [22, 26].

Advances in 3D printing technology allow for the creation of anatomically accurate brain tissue models. Various materials, including soft elastomers, can be 3D printed to simulate brain tissue [35]. Phantom tissues are commercially available materials designed to replicate the acoustic and mechanical properties of biological tissues, including brain tissue. These may be used in medical imaging applications. Researchers may combine different polymeric materials to create blends that offer a more comprehensive representation of brain tissue properties. This allows for the customization of material characteristics. PMMA is a transparent thermoplastic that has been used in the construction of transparent brain models for educational purposes. While it may not replicate the mechanical properties of brain tissue accurately, it allows for visual representation [10, 14]. Nanocomposites, incorporating nanoparticles such as carbon nanotubes, can be used to enhance the electrical conductivity of the simulant, mimicking the conductive properties of neurons. Polymers with inherent electrical conductivity, such as polypyrrole and poly(3,4-ethylenedioxythiophene) (PEDOT), can be used to replicate the electrical properties of neurons [36]. Some researchers use biological materials like agarose, collagen, or decellularized extracellular matrix (ECM) scaffolds to create more biologically realistic brain tissue simulants [22].

The choice of material depends on the specific requirements of the simulation, such as the need for electrical conductivity, biocompatibility, or imaging compatibility (see Fig. 5.2). Often, a combination of materials is used to achieve a more comprehensive representation of brain tissue properties. Additionally, advancements

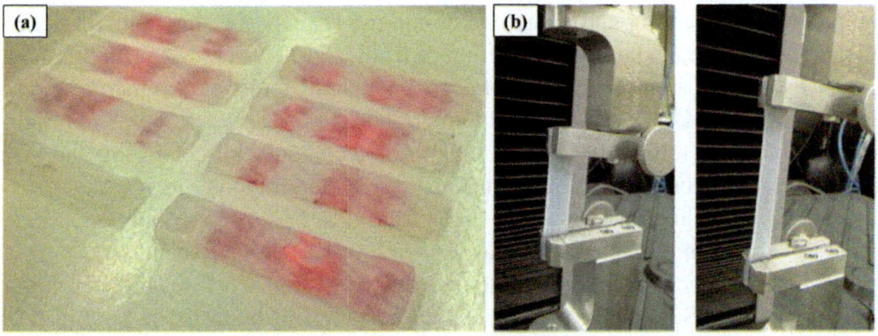

Fig. 5.2 **a** Silicone-based brain tissue simulants, **b** tested under tensile loading [8]

in materials science and tissue engineering continually contribute to the development of more realistic brain tissue simulants.

5.3 Application Areas of Brain Tissue Simulants

Brain tissue simulants, also known as brain mimics or brain models, are artificial materials designed to replicate certain properties of the human brain. These simulants serve various purposes in scientific research, medical applications, and technology development. Brain tissue simulants are crucial for studying brain function, neural processes, and the effects of drugs or treatments. Researchers use these models to test hypotheses, conduct experiments, and gain insights into the complex workings of the brain without the ethical and practical challenges associated with human or animal studies [27]. Simulants are used in medical training programs to simulate realistic conditions for surgeons and medical professionals. They provide a safe and controlled environment for practicing surgical procedures, particularly those involving the brain. This helps in refining surgical skills and improving outcomes for patients. Brain tissue simulants play a vital role in drug discovery and testing. Researchers use these models to assess the potential impact of drugs on the brain, identify potential side effects, and evaluate the efficacy of new treatments. This helps in streamlining the drug development process and minimizing risks associated with human trials [37–40].

Simulants are employed in the development and testing of neuroprosthetics and brain-machine interfaces. These technologies aim to interface with the brain to restore or enhance sensory, motor, or cognitive functions. Simulants help researchers understand how these devices interact with neural tissue and improve their safety and effectiveness. Brain tissue simulants are used in educational settings to teach students about neurobiology, neuroscience, and related fields. They provide a hands-on and tangible way for students to explore and understand the structure and function of the brain without the need for specialized facilities or live specimens. Students in biomedical engineering and related fields can use brain tissue simulants for experimentation, testing, and design validation of medical devices [29, 41, 42]. Simulants can be used to develop and calibrate diagnostic imaging tools, such as magnetic resonance imaging (MRI) and computed tomography (CT) scans. This helps in refining the accuracy and reliability of these diagnostic techniques in identifying and visualizing brain abnormalities. Simulants aid researchers in studying various neurological disorders and diseases, such as Alzheimer's, Parkinson's, and epilepsy. By creating models that mimic specific aspects of these conditions, scientists can investigate disease mechanisms, test potential therapies, and work toward developing treatments [24]. Simulating the mechanical response of brain tissue is crucial for studying traumatic brain injuries (TBIs) and understanding the impact of different forces on the brain. Such precisely developed tissue simulants would be beneficial and allow researchers to investigate the biomechanics of the brain, including its response to forces, stretching, and deformation. Brain tissue simulants offer an alternative to animal testing in certain experiments and validations, aligning with ethical considerations and regulations that encourage the reduction of animal use in research. Also, the simulants contribute to the standardization of testing procedures and the validation of medical devices, ensuring consistency and reproducibility in research and development. Researchers can customize brain tissue simulants to replicate specific mechanical, electrical, or thermal properties relevant to their study or application. The primary advantage of the simulants is that they allow reproducible experiments, ensuring that studies can be replicated across different laboratories or research settings [22, 24].

In summary, brain tissue simulants are invaluable tools that contribute to advancements in neuroscience, medicine, and technology. They provide a controlled and ethical means for researchers and practitioners to explore the intricacies of the brain and develop innovative solutions for a variety of applications.

References

1. Budday S, Sommer G, Birkl C, Langkammer C, Haybaeck J, Kohnert J et al (2017) Mechanical characterization of human brain tissue. Acta Biomater 48:319–340. https://doi.org/10.1016/j.actbio.2016.10.036
2. Budday S, Sarem M, Starck L, Sommer G, Pfefferle J, Phunchago N et al (2020) Towards microstructure-informed material models for human brain tissue. Acta Biomater 104:53–65. https://doi.org/10.1016/j.actbio.2019.12.030
3. Prange MT, Margulies SS (2002) Regional, directional, and age-dependent properties of the brain undergoing large deformation. J Biomech Eng 124:244–252. https://doi.org/10.1115/1.1449907
4. Takhounts EG, Crandall JR, Darvish K (2003) On the Importance of Nonlinearity of Brain Tissue under Large Deformations. SAE Tech. Pap., vol 2003-Octob, SAE Internationa. https://doi.org/10.4271/2003-22-0005
5. Sack I, Streitberger K-J, Krefting D, Paul F, Braun J (2011) The influence of physiological aging and atrophy on brain viscoelastic properties in humans. PLoS ONE 6:e23451. https://doi.org/10.1371/journal.pone.0023451

6. Biran R, Martin DC, Tresco PA (2005) Neuronal cell loss accompanies the brain tissue response to chronically implanted silicon microelectrode arrays. Exp Neurol 195:115–126. https://doi.org/10.1016/J.EXPNEUROL.2005.04.020
7. Green MA, Bilston LE, Sinkus R (2008) In vivo brain viscoelastic properties measured by magnetic resonance elastography. NMR Biomed 21:755–764. https://doi.org/10.1002/nbm.1254
8. Chanda A, Callaway C, Clifton C, Unnikrishnan V (2018) Biofidelic human brain tissue surrogates. Mech Adv Mater Struct 25:1335–1341. https://doi.org/10.1080/15376494.2016.1143749
9. Polikov VS, Tresco PA, Reichert WM (2005) Response of brain tissue to chronically implanted neural electrodes. J Neurosci Methods 148:1–18. https://doi.org/10.1016/J.JNEUMETH.2005.08.015
10. Cloots RJH, Van Dommelen JAW, Kleiven S, Geers MGD (2013) Multi-scale mechanics of traumatic brain injury: predicting axonal strains from head loads. Biomech Model Mechanobiol 12:137–150. https://doi.org/10.1007/S10237-012-0387-6/METRICS
11. Rashid B, Destrade M, Gilchrist MD (2012) Mechanical characterization of brain tissue in compression at dynamic strain rates. J Mech Behav Biomed Mater 10:23–38. https://doi.org/10.1016/j.jmbbm.2012.01.022
12. Yeung J, Jugé L, Hatt A, Bilston LE (2019) Paediatric brain tissue properties measured with magnetic resonance elastography. Biomech Model Mechanobiol 18:1497–1505. https://doi.org/10.1007/s10237-019-01157-x
13. Zhu Z, Jiang C, Jiang H (2019) A visco-hyperelastic model of brain tissue incorporating both tension/compression asymmetry and volume compressibility. Acta Mech 230:2125–2135. https://doi.org/10.1007/s00707-019-02383-1
14. Pervin F, Chen WW (2011) Mechanically similar gel simulants for brain tissues. Conf Proc Soc Exp Mech Ser 1:9–13. https://doi.org/10.1007/978-1-4419-8228-5_3/COVER
15. Gefen A, Margulies SS (2004) Are in vivo and in situ brain tissues mechanically similar? J Biomech 37:1339–1352. https://doi.org/10.1016/J.JBIOMECH.2003.12.032
16. Bilston LE (2011) Brain tissue mechanical properties. Springer, New York, NY, pp 69–89. https://doi.org/10.1007/978-1-4419-9997-9_4
17. Huang X, Chafi H, Matthews KL, Carmichael O, Li T, Miao Q et al (2019) Magnetic resonance elastography of the brain: a study of feasibility and reproducibility using an ergonomic pillow-like passive driver. Magn Reson Imaging 59:68–76. https://doi.org/10.1016/j.mri.2019.03.009
18. Budday S, Nay R, de Rooij R, Steinmann P, Wyrobek T, Ovaert TC et al (2015) Mechanical properties of gray and white matter brain tissue by indentation. J Mech Behav Biomed Mater 46:318–330. https://doi.org/10.1016/j.jmbbm.2015.02.024
19. Shuck LZ, Advani SH (1972) Rheologioal response of human brain tissue in shear. J Fluids Eng Trans ASME 94:905–911. https://doi.org/10.1115/1.3425588
20. Chatelin S, Constantinesco A, Willinger R (2010) Fifty years of brain tissue mechanical testing: from in vitro to in vivo investigations. Biorheology 47:255–276. https://doi.org/10.3233/BIR-2010-0576
21. Miller K, Chinzei K (2002) Mechanical properties of brain tissue in tension. J Biomech 35:483–490. https://doi.org/10.1016/S0021-9290(01)00234-2
22. Singh G, Chanda A (2021) Mechanical properties of whole-body soft human tissues: a review. Biomed Mater 16:062004. https://doi.org/10.1088/1748-605X/AC2B7A
23. Zhang W, Liu L, Xiong Y, Liu Y, Yu S, Wu C, et al (2018) Effect of in vitro storage duration on measured mechanical properties of brain tissue. Sci Rep 8. https://doi.org/10.1038/s41598-018-19687-2
24. Singh G, Chanda A (2023) Development and mechanical characterization of artificial surrogates for brain tissues. Biomed Eng Adv 5:100084. https://doi.org/10.1016/J.BEA.2023.100084
25. Chanda A, Callaway C (2018) Tissue anisotropy modeling using soft composite materials. Appl Bionics Biomech 2018. https://doi.org/10.1155/2018/4838157
26. Chanda A, Singh G (2023) Tissues in functional organs—low stiffness. Mater Horizons From Nat to Nanomater:33–48. https://doi.org/10.1007/978-981-99-2225-3_4/COVER

27. Chanda A, Singh G (2023) Applications, challenges, and future opportunities. Mater Horizons From Nat to Nanomater:85–92. https://doi.org/10.1007/978-981-99-2225-3_8/COVER
28. Singh G, Gupta V, Chanda A (2022) Artificial skin with varying biomechanical properties. Mater Today Proc 62:3162–3166. https://doi.org/10.1016/J.MATPR.2022.03.433
29. Gupta V, Singla R, Singh G, Chanda A (2023) Development of soft composite based anisotropic synthetic skin for biomechanical testing. Fibers 11:55. https://doi.org/10.3390/FIB11060055
30. Makode S, Singh G, Chanda A (2021) Development of novel anisotropic skin simulants. Phys Scr 96:125019. https://doi.org/10.1088/1402-4896/AC2EFD
31. Gupta V, Singh G, Gupta S, Chanda A (2023) Expansion potential of auxetic prosthetic skin grafts: a review. Eng Res Express 5:022003. https://doi.org/10.1088/2631-8695/ACCFE5
32. Chanda A (2018) Biomechanical modeling of human skin tissue surrogates. Biomimetics 3:18. https://doi.org/10.3390/BIOMIMETICS3030018
33. Singh G, Chanda A (2023) Biofidelic tongue and tonsils tissue surrogates. Mater Horizons From Nat to Nanomater; Part F1471:159–70. https://doi.org/10.1007/978-981-99-5064-5_10/COVER
34. Levental I, Georges PC, Janmey PA (2007) Soft biological materials and their impact on cell function. Soft Matter 3:299–306. https://doi.org/10.1039/B610522J
35. Chhikara K, Singh G, Gupta S, Chanda A (2022) Progress of additive manufacturing in fabrication of foot orthoses for diabetic patients: a review. Ann 3D Print Med 8:100085. https://doi.org/10.1016/J.STLM.2022.100085
36. Chanda A, Unnikrishnan V, Lackey K, Robbins J (2020) Biofidelic conductive soft tissue surrogates. Int J Polym Mater Polym Biomater 69:127–135. https://doi.org/10.1080/00914037.2018.1552856
37. Chanda A, Singh G (2023) Introduction to human tissues. Mater Horizons From Nat to Nanomater:1–12. https://doi.org/10.1007/978-981-99-2225-3_1/COVER
38. Chanda A, Unnikrishnan V, Flynn Z, Lackey K (2017) Experimental study on tissue phantoms to understand the effect of injury and suturing on human skin mechanical properties. Proc Inst Mech Eng Part H J Eng Med 231:80–91. https://doi.org/10.1177/0954411916679438
39. Singh G, Chanda A (2023) Development and biomechanical testing of artificial surrogates for vaginal tissue. Adv Mater Process Technol. https://doi.org/10.1080/2374068X.2023.2198837
40. Singh G, Chanda A (2023) Biofidelic gallbladder tissue surrogates. Adv Mater Process Technol. https://doi.org/10.1080/2374068X.2023.2198835
41. Gupta V, Singh G, Chanda A (2023) Development of novel hierarchical designs for skin graft simulants with high expansion potential. Biomed Phys Eng Express 9:035024. https://doi.org/10.1088/2057-1976/ACC661
42. Gupta V, Singh G, Chanda A (2023) High Expansion Auxetic Skin Graft Simulants for Severe Burn Injury Mitigation. Eur Burn J 4:108–20. https://doi.org/10.3390/EBJ4010011

Chapter 6
Lung Tissue Simulants

Human lung tissue is a complex and highly specialized organ responsible for the exchange of oxygen and carbon dioxide, a process essential for respiration. The lungs are vital components of the respiratory system, facilitating the exchange of gases between the air and the bloodstream [1, 2]. The lungs are a pair of large, sponge-like organs that are located in the thorax, lateral to the heart and superior to the diaphragm. Compared to the right lung, the left lung is considerably smaller and has two lobes, while the right lung has three lobes [3, 4].

6.1 Anatomy of the Lungs

- *Lobes and Lobules*: The human lungs are paired organs, with the right lung typically having three lobes (upper, middle, and lower lobes) and the left lung having two lobes (upper and lower lobes). Each lobe is further divided into smaller units called lobules [3].
- *Bronchial Tree*: The bronchial tree comprises a series of branching airways that lead to and within the lungs. It starts with the trachea, which divides into the main bronchi (right and left) that enter each lung. The bronchi further divide into smaller bronchi and bronchioles.
- *Bronchioles*: Bronchioles are smaller airways that extend deep into the lungs. They eventually lead to structures called alveoli.
- *Alveoli*: Alveoli are tiny, balloon-like structures at the end of the bronchioles. These are the primary sites for gas exchange. The walls of the alveoli are thin and surrounded by a network of capillaries, allowing oxygen to diffuse into the bloodstream and carbon dioxide to move out [4, 5].

A. Chanda and G. Singh, *Soft Tissue Simulants*, Biomedical Materials for Multi-functional Applications, https://doi.org/10.1007/978-981-97-3060-5_6

- *Pleura*: The lungs are enclosed in a double-layered membrane called the pleura. The visceral pleura covers the lung surface, while the parietal pleura lines the chest cavity. The space between these layers is the pleural cavity, filled with a small amount of lubricating fluid, which reduces friction during breathing [6].

6.2 Microscopic Structure of Lung Tissue

- *Epithelial Cells*: The respiratory epithelium lines the airways and is composed of various cell types, including ciliated cells and goblet cells. Ciliated cells have hair-like structures (cilia) that move mucus and trapped particles upward toward the throat. Goblet cells produce mucus to trap and moisten inhaled air [7].
- *Alveolar Type I and Type II Cells*: Type I cells form the majority of the alveolar wall and are involved in gas exchange. Type II cells secrete surfactant, a substance that reduces surface tension in the alveoli, preventing their collapse during exhalation [8, 9].
- *Capillaries*: Capillaries surround the alveoli, facilitating the exchange of oxygen and carbon dioxide with the bloodstream. This process is known as pulmonary gas exchange [2].
- *Connective Tissue*: Connective tissue provides structural support to the lungs, helping to maintain their shape and elasticity. Elastic fibers allow the lungs to expand during inhalation and recoil during exhalation [10].

6.3 Functions of Lung Tissue

- *Gas Exchange*: The primary function of lung tissue is to facilitate the exchange of oxygen and carbon dioxide between the air in the lungs and the bloodstream. Oxygen is taken up by red blood cells in the capillaries, and carbon dioxide is released for exhalation [8].
- *Ventilation*: Lung tissue plays a crucial role in the process of ventilation, which involves the movement of air into and out of the lungs. This process is essential for maintaining adequate levels of oxygen and removing carbon dioxide [8, 9].
- *Surfactant Production*: Type II cells produce surfactant, a substance that reduces surface tension in the alveoli. This prevents the collapse of alveoli and facilitates the ease of breathing [6].
- *Immune Defense*: The respiratory epithelium, along with mucus-producing cells, provides a barrier against inhaled pathogens and particles. Cilia help in the removal of mucus and trapped particles, contributing to the immune defense of the respiratory system [6].
- *pH Regulation*: The lungs play a role in regulating the pH of the blood by influencing the levels of carbon dioxide, a key factor in maintaining the body's acid–base balance [6].

6.4 Mechanical Aspects of Human Lung Tissue

The mechanical aspects of human lung tissue involve its response to forces, deformations, and pressures, which are crucial for understanding respiratory function, biomechanics, and the behavior of the lungs under various conditions [11].

Elasticity: Elasticity refers to the ability of lung tissue to deform reversibly under an applied force and return to its original shape when the force is removed. The elastic properties of lung tissue allow it to stretch during inhalation and recoil during exhalation. Elastic fibers in the connective tissue provide the necessary recoil [12].

Compliance: Compliance is a measure of the ease with which the lungs expand in response to changes in pressure. Compliance is crucial for normal breathing. Healthy lung tissue has optimal compliance, allowing for efficient inhalation and exhalation. Diseases like fibrosis can reduce compliance [1, 13].

Surface Tension: Surface tension is the force at the air–liquid interface that tends to reduce the surface area of the liquid. Surface tension in the alveoli is reduced by surfactant, preventing alveolar collapse during exhalation. This is essential for maintaining lung compliance [6].

Viscoelasticity: Viscoelasticity combines the characteristics of viscosity and elasticity, indicating how lung tissue deforms over time in response to a force. Lung tissue exhibits viscoelastic behavior, influencing the time-dependent aspects of breathing and the damping of oscillations during respiratory cycles [11, 14].

Hysteresis: Hysteresis refers to the lag between the inflation and deflation curves on a pressure–volume graph during the respiratory cycle. Hysteresis is observed due to energy dissipation within the lung tissue during breathing. It is related to the viscoelastic properties of the lung [2].

Airway Resistance: Airway resistance is the opposition encountered by airflow in the respiratory passages. Resistance is influenced by factors such as the diameter of the airways and the viscosity of the air. Conditions like asthma or chronic obstructive pulmonary disease (COPD) can increase airway resistance [15, 16].

Stress and Strain: Stress is the force applied per unit area, and strain is the resulting deformation. These concepts are used to describe how lung tissue responds to mechanical loads. Studying stress and strain helps understand the mechanical properties of lung tissue under different conditions and during respiratory cycles [17, 18].

Anisotropy: Anisotropy refers to the direction-dependent mechanical properties of a material. Lung tissue exhibits anisotropic behavior, with mechanical properties varying in different directions. This is important for accurately modeling and understanding lung mechanics [19, 20].

Pulmonary Compliance and Diseases: The overall compliance of the respiratory system, including the lung and chest wall. Diseases such as pulmonary fibrosis, emphysema, and certain forms of lung disease can alter pulmonary compliance, affecting the mechanical properties of the lung tissue [21].

Biomechanical Models: Biomechanical models and computational simulations are used to study the mechanical behavior of lung tissue under various conditions, contributing to research on respiratory mechanics [22, 23].

Understanding the mechanical aspects of human lung tissue is essential for clinicians, researchers, and engineers involved in respiratory medicine, biomechanics, and medical device design. It provides insights into normal lung function, the impact of diseases, and the development of interventions to maintain or restore respiratory health. Research in this area contributes to advancements in the diagnosis and treatment of respiratory conditions.

6.5 Traditional Lung Tissues for Testing

Traditional lung tissues for testing in research and medical contexts typically involve the use of animal models and post-mortem human tissues. These approaches have been widely utilized for studying lung physiology, and pathology, and for testing various medical interventions [6, 24, 25].

6.5.1 Animal Models

- *Rats and Mice*: Small rodents are commonly used for respiratory research due to their size, cost-effectiveness, and availability. Transgenic and genetically modified mice are valuable for studying specific aspects of lung function and disease [6].
- *Pigs*: Pigs have a respiratory system that is similar to humans, making them suitable for studies related to respiratory physiology, mechanics, and disease [6].
- *Non-Human Primates*: Primates, such as macaques, share physiological similarities with humans and are used in studies that require a closer approximation to human lung function [6].

6.5.2 Post-mortem Human Tissues

Tissues obtained from deceased donors are used for various research purposes. These tissues may be procured from organ donors or cadavers. The second type are lung biopsy samples, where tissue samples obtained through lung biopsies during diagnostic or therapeutic procedures can be used for research purposes [6].

6.5.3 Precision-Cut Lung Slices

Precision-cut lung slices are thin sections of lung tissue that are prepared and maintained under controlled conditions. These slices retain the native architecture and allow for ex vivo studies, including drug testing, toxicity assessments, and investigating lung responses to various stimuli [6].

6.5.4 Humanized Mouse Models

Humanized mouse models involve introducing human lung tissue or cells into mice to create a more human-like environment. This approach is useful for studying interactions between the human immune system and lung tissues. Some institutions maintain biobanks that store human lung tissues for research purposes. These banks facilitate the availability of diverse tissue samples for investigators [1, 2, 9].

These traditional lung tissues and models have been valuable in advancing our understanding of lung biology, disease mechanisms, and therapeutic interventions. However, it's important to note that ethical considerations, variability between species, and limitations in recapitulating complex in vivo conditions are challenges associated with the use of these models. Advanced technologies, such as organoids and bioengineered lung tissues, are emerging as alternative approaches that aim to overcome some of these limitations. Only a few research have explored human lung tissue experimentally since cadaveric lung tissue is not easily accessible [18, 26, 27]. Recent research has shown a significant interest in the computational modeling of lung tissue, where various imaging techniques (such as CT scans and MRIs) are being utilized to create a realistic lung model (Fig. 6.1) [28].

6.6 Lung Tissue Simulants

Lung tissue simulants are synthetic materials designed to mimic the mechanical and physiological properties of natural lung tissue. These simulants serve as valuable tools for research, medical training, and the development and testing of medical devices. The use of lung tissue simulants provides a controlled and reproducible environment, reducing the reliance on traditional models and ethical considerations associated with live animal or human tissues. Creating realistic lung tissue simulants involves selecting materials that mimic the structural, mechanical, and physiological properties of native lung tissue [1, 14]. Silicone rubber is a popular choice due to its excellent elasticity and durability. It can be molded to replicate the soft and stretchable nature of lung tissue. Silicone rubber provides realistic tactile feedback, making it suitable for surgical training and medical device testing. Latex rubber is another elastic material that can be used to simulate the texture and feel of lung

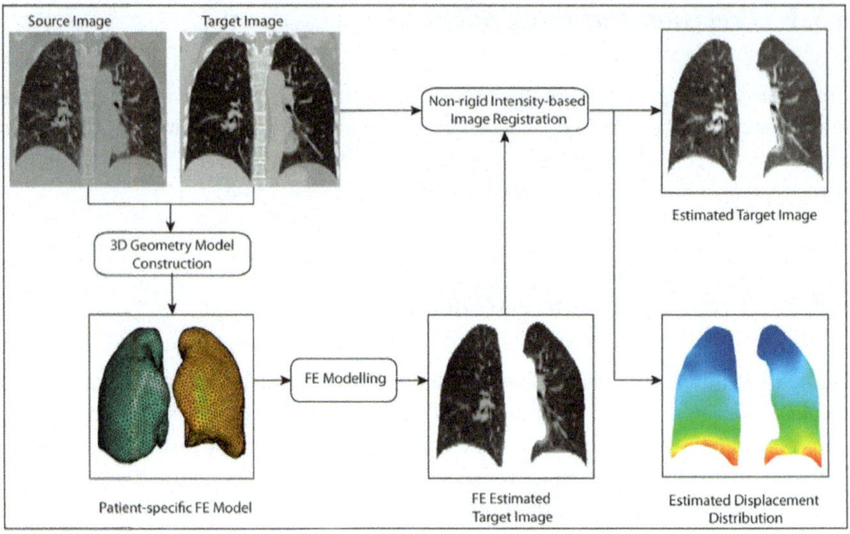

Fig. 6.1 Patient-specific biomechanical modeling of lung tissue [28]

tissue. It is often employed in medical training scenarios for procedures involving palpation and manipulation of soft tissues [21, 26, 29–31]. Agarose and Alginate are two commonly used hydrogels to mimic soft tissues. These hydrogels are used to simulate the gel-like consistency of lung tissue. They can be formulated to have similar mechanical properties and are often employed in 3D printing applications. Alginate, in particular, is suitable for creating vascularized structures within the simulated lung. Another type of hydrogel, i.e., polyacrylamide hydrogels can be tuned to replicate specific mechanical properties of lung tissue. They are used in research settings to study cellular responses and interactions within a controlled environment [32–34]. Polyurethane foam is a lightweight and porous material, which can be shaped to simulate the spongy texture of lung parenchyma. Polyurethane foam is commonly used in medical imaging phantoms, providing a realistic medium for ultrasound and CT imaging studies. Latex foam, similar to rubber, is used to create lung models with compressible and resilient properties. It is often utilized in medical training scenarios for procedures involving incisions and suturing. Collagen, a key component of the extracellular matrix in the lungs, can be used to create tissue-like scaffolds. These matrices support cell growth and migration, making them valuable for studies in lung tissue engineering and regeneration [6, 35, 36].

Poly(lactic-co-glycolic acid) (PLGA) is a biodegradable polymer used in tissue engineering. It can be shaped into scaffolds that provide mechanical support and gradually degrade as new tissue forms. PEG hydrogels are biocompatible and can be customized to mimic the mechanical properties of lung tissue. They are often employed in microfluidic devices to study lung-related phenomena at the cellular level [36, 37]. Bioinks formulated with a combination of hydrogels, cells, and other bioactive components are used in 3D bioprinting to create anatomically accurate lung tissue models. These models can replicate the microarchitecture and complexity of native lung tissue [1, 2, 13].

Lung tissue simulants play a crucial role in advancing medical education, research, and the development of medical technologies. They offer a controlled and reproducible platform for various applications, allowing researchers and medical professionals to study lung function, test devices, and explore new treatment approaches in a more ethical and controlled manner [13]. It's important to consider the specific application and requirements when selecting materials for lung tissue simulants. The ideal simulant should closely mimic the properties of native lung tissue to ensure accurate representation in various medical and research contexts. Ongoing advancements in material science and tissue engineering continue to contribute to the development of more realistic lung tissue simulants. In recent years, across all these discussed materials, silicone rubber and 3D printing are the prominently used materials for creating the simulants for lung tissue [6].

6.6.1 Silicone Rubber

The primary reason for using silicone is its advantage over the other materials discussed above. Silicone rubber has excellent elasticity and flexibility with durable and long-lasting implementation. It is effective in mimicking soft tissues such as lung tissue and can be customized for specific mechanical properties. Also, the fabrication process is simple and does not need any specific training or practice [38, 39]. Silicone rubber is typically available in liquid form. It is mixed with a curing agent to initiate the crosslinking process and the liquid silicone rubber is poured into a mold designed to replicate the anatomical structure of the lung. The mold may be based on medical imaging data or anatomical models. Then, the material is allowed to cure and solidify within the mold. The curing time varies depending on the specific silicone formulation used. Once cured, the simulated lung is carefully removed from the mold. The resulting structure exhibits the desired elasticity and flexibility [26, 31].

Elastomeric lung simulants are valuable for training medical professionals in procedures such as bronchoscopy, thoracentesis, and lung palpation. These simulants are used in research settings to test medical devices, study biomechanics, and develop surgical techniques. Simulated lungs are employed in patient education programs to visually and tactilely represent anatomical structures and conditions [6, 25].

6.6.2 3D Bioprinting

Fabricating lung tissue simulants using 3D printing involves creating artificial structures that mimic the anatomy of real lung tissue. 3D printing, also known as additive manufacturing, allows for the precise layer-by-layer deposition of materials to build complex structures. The first step is the selection of the printing materials. Biocompatible polymer materials that mimic the mechanical and structural properties of lung tissue such as polylactic acid (PLA), polyethylene glycol (PEG), and polyurethane are commonly used due to their biocompatibility and suitability for 3D printing [2, 13]. To accurately replicate lung anatomy, starting with medical imaging data such as CT (computed tomography) or MRI (magnetic resonance imaging) scans are important. These scans can be used to create a 3D digital model of the lung. Thereafter, CAD software can be used to convert the imaging data into a digital 3D model. The digital model serves as a blueprint for the 3D printer to create the physical structure [13].

The printing process which includes layer-by-layer deposition by a 3D printer, deposit material layer by layer according to the digital model. This layering process is crucial for creating intricate structures that resemble the complexity of lung tissue. Depending on the complexity of the lung tissue model, support structures may be required during the printing process to maintain the integrity of overhanging features. After printing, remove any support structures used during the printing process [13, 31, 40]. This may involve mechanical or chemical processes, depending on the printing material. Depending on the printing resolution, additional post-processing steps like sanding or polishing may be required to achieve a smoother surface texture. The developed 3D-printed lung tissue simulant can be used for research, drug testing, and medical education purposes [6] (Fig. 6.2).

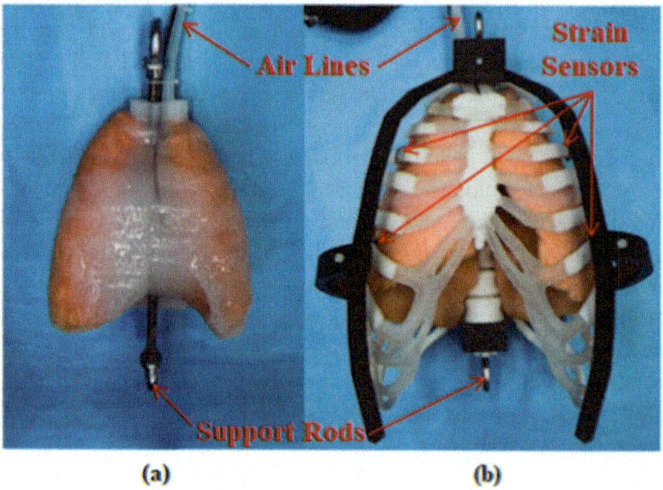

Fig. 6.2 Printed to-scale artificial adult human lungs that are **a** bare and **b** with the ribcage [41]

6.7 Application Areas of Lung Tissue Simulants

Lung tissue simulants find applications in various fields, including medical education, research, and the development of medical devices. These synthetic materials provide a controlled and reproducible environment for studying lung function, testing devices, and simulating physiological conditions. Lung tissue simulants are utilized in medical training for surgeons to practice procedures such as lung biopsies, lobectomies, and other respiratory interventions in a controlled and risk-free environment [42, 43]. These simulants serve as educational tools in medical schools and training programs, allowing students to gain hands-on experience with realistic lung models. Lung tissue simulants provide a platform for modeling various respiratory diseases, including chronic obstructive pulmonary disease (COPD), asthma, and lung cancer. Researchers can study disease progression, test potential therapeutics, and explore underlying mechanisms. Patient-specific lung tissue simulants allow for the testing of personalized treatment strategies, considering individual variations in drug responses and disease characteristics [6, 44, 45].

Lung tissue simulants can be employed to test and optimize biosensors for detecting specific biomarkers associated with lung diseases. This contributes to the development of diagnostic tools and monitoring devices. Simulants allow researchers to study cell behavior, migration, and interaction within a 3D microenvironment, providing insights into cellular responses to stimuli and environmental cues. Lung tissue simulants aid in understanding the interaction between transplanted organs and the recipient's immune system [8, 9, 40]. This knowledge is vital for improving transplantation outcomes and reducing rejection risks. Lung tissue simulants contribute to the advancement of tissue engineering approaches, helping researchers design and optimize scaffolds, cell seeding methods, and culture conditions for developing functional lung tissues for transplantation.

References

1. Faffe DS, Silva GH, Kurtz PMP, Negri EM, Capelozzi VL, Rocco PRM et al (2001) Lung tissue mechanics and extracellular matrix composition in a murine model of silicosis. J Appl Physiol 90:1400–1406. https://doi.org/10.1152/JAPPL.2001.90.4.1400/ASSET/IMAGES/LARGE/DG0410582004.JPEG
2. Ito S, Ingenito EP, Brewer KK, Black LD, Parameswaran H, Lutchen KR et al (2005) Mechanics, nonlinearity, and failure strength of lung tissue in a mouse model of emphysema: possible role of collagen remodeling. J Appl Physiol 98:503–511. https://doi.org/10.1152/JAPPLPHYSIOL.00590.2004/ASSET/IMAGES/LARGE/ZDG0020536060008.JPEG
3. Bou Jawde S, Takahashi A, Bates JHT, Suki B (2020) An analytical model for estimating alveolar wall elastic moduli from lung tissue uniaxial stress-strain curves. Front Physiol 11:121. https://doi.org/10.3389/fphys.2020.00121
4. Andrikakou P, Vickraman K, Arora H (2016) On the behaviour of lung tissue under tension and compression. Sci Rep 6:1–10. https://doi.org/10.1038/srep36642
5. Mercer RR, Crapo JD (1990) Spatial distribution of collagen and elastin fibers in the lungs. J Appl Physiol 69:756–765. https://doi.org/10.1152/jappl.1990.69.2.756

6. Singh G, Chanda A (2021) Mechanical properties of whole-body soft human tissues: a review. Biomed Mater 16:062004. https://doi.org/10.1088/1748-605X/AC2B7A
7. Mercer RR, Russell ML, Crapo JD (1994) Alveolar septal structure in different species. J Appl Physiol 77:1060–1066. https://doi.org/10.1152/jappl.1994.77.3.1060
8. Matsunashi A, Nagata K, Morimoto T, Tomii K (2023) Mechanical ventilation for acute exacerbation of fibrosing interstitial lung diseases. Respir Investig 61:306–313. https://doi.org/10.1016/J.RESINV.2023.01.008
9. Olasveengen TM, Skåre C, Skjerven-Martinsen M, Hoff-Olsen P, Kramer-Johansen J, Hoff Nordum F et al (2024) Lung tissue injury and hemodynamic effects of ventilations synchronized or unsynchronized to continuous chest compressions in a porcine cardiac arrest model. Resusc Plus 17:100530. https://doi.org/10.1016/J.RESPLU.2023.100530
10. Chanda A, Singh G (2023) Muscles and connective tissues. In: Materials horizons: from nature to nanomaterials, pp 25–32. https://doi.org/10.1007/978-981-99-2225-3_3/COVER
11. Sebag SC, Bastarache JA, Ware LB (2013) Mechanical stretch inhibits lipopolysaccharide-induced keratinocyte-derived chemokine and tissue factor expression while increasing procoagulant activity in murine lung epithelial cells. J Biol Chem 288:7875–7884. https://doi.org/10.1074/JBC.M112.403220
12. Panzetta V, Musella I, Rapa I, Volante M, Netti PA, Fusco S (2017) Mechanical phenotyping of cells and extracellular matrix as grade and stage markers of lung tumor tissues. Acta Biomater 57:334–341. https://doi.org/10.1016/J.ACTBIO.2017.05.002
13. Karamchand L, Makeiff D, Gao Y, Azyat K, Serpe MJ, Kulka M (2023) Biomaterial inks and bioinks for fabricating 3D biomimetic lung tissue: a delicate balancing act between biocompatibility and mechanical printability. Bioprinting 29:e00255. https://doi.org/10.1016/J.BPRINT.2022.E00255
14. Shirani A, Ganji F, Golmohammadi M, Hashemi SM, Mozafari M, Amoabediny G et al (2021) Cross-linked acellular lung for application in tissue engineering: effects on biocompatibility, mechanical properties and immunological responses. Mater Sci Eng C 122:111938. https://doi.org/10.1016/J.MSEC.2021.111938
15. Singh G, Chanda A (2023) Biofidelic tongue and tonsils tissue surrogates. In: Materials horizons: from nature to nanomaterials, Part F1471, pp 159–70. https://doi.org/10.1007/978-981-99-5064-5_10/COVER
16. Chanda A, Singh G (2023) Tissues in functional organs—low stiffness. In: Materials horizons: from nature to nanomaterials, pp 33–48. https://doi.org/10.1007/978-981-99-2225-3_4/COVER
17. Singh G, Gupta V, Chanda A (2022) Artificial skin with varying biomechanical properties. Mater Today Proc 62:3162–3166. https://doi.org/10.1016/J.MATPR.2022.03.433
18. Singh G, Chanda A (2022) Biomechanical modeling of progressive wound healing: a computational study. Biomed Eng Adv 4:100055. https://doi.org/10.1016/J.BEA.2022.100055
19. Gupta V, Singla R, Singh G, Chanda A (2023) Development of soft composite based anisotropic synthetic skin for biomechanical testing. Fibers 11:55. https://doi.org/10.3390/FIB11060055
20. Chanda A, Singh G (2023) Hyperelastic models for anisotropic tissue characterization. In: Materials horizons: from nature to nanomaterials, pp 73–83. https://doi.org/10.1007/978-981-99-2225-3_7/COVER
21. Makode S, Singh G, Chanda A (2021) Development of novel anisotropic skin simulants. Phys Scr 96:125019. https://doi.org/10.1088/1402-4896/AC2EFD
22. Singh G, Chanda A (2023) Biofidelic gallbladder tissue surrogates. Adv Mater Process Technol. https://doi.org/10.1080/2374068X.2023.2198835
23. Chanda A, Unnikrishnan V, Lackey K, Robbins J (2020) Biofidelic conductive soft tissue surrogates. Int J Polym Mater Polym Biomater 69:127–135. https://doi.org/10.1080/00914037.2018.1552856
24. Gupta V, Singh G, Chanda A (2023) Modeling of metamaterial based incision patterns for generating high expansions in skin grafts. Clin Biomech 110:106118. https://doi.org/10.1016/J.CLINBIOMECH.2023.106118

25. Chanda A, Singh G (2023) Introduction to human tissues. In: Materials horizons: from nature to nanomaterials, pp 1–12. https://doi.org/10.1007/978-981-99-2225-3_1/COVER
26. Gupta V, Singh G, Chanda A (2022) Development and testing of skin grafts models with varying slit orientations. Mater Today Proc 62:3462–3467. https://doi.org/10.1016/J.MATPR.2022.04.282
27. Gupta V, Singh G, Gupta S, Chanda A (2023) Expansion potential of auxetic prosthetic skin grafts: a review. Eng Res Express 5:022003. https://doi.org/10.1088/2631-8695/ACCFE5
28. Han L, Dong H, McClelland JR, Han L, Hawkes DJ, Barratt DC (2017) A hybrid patient-specific biomechanical model based image registration method for the motion estimation of lungs. Med Image Anal 39:87–100. https://doi.org/10.1016/j.media.2017.04.003
29. Singh G, Chanda A (2023) Development and mechanical characterization of artificial surrogates for brain tissues. Biomed Eng Adv 5:100084. https://doi.org/10.1016/J.BEA.2023.100084
30. Gupta V, Gupta S, Chanda A (2023) Development of an ultra-low-cost planar biaxial tester for soft tissue characterization. Biomed Phys Eng Express 9:025011. https://doi.org/10.1088/2057-1976/ACB940
31. Gupta V, Singh G, Chanda A (2023) Development of hierarchical auxetic skin graft simulants with high expansion potential. Biomed Eng Adv 5:100087. https://doi.org/10.1016/J.BEA.2023.100087
32. Vikatmaa P, Juutilainen V, Kuukasjärvi P, Malmivaara A (2008) Negative pressure wound therapy: a systematic review on effectiveness and safety. Eur J Vasc Endovasc Surg 36:438–448. https://doi.org/10.1016/J.EJVS.2008.06.010
33. Gupta V, Singh G, Chanda A (2023) High expansion auxetic skin graft simulants for severe burn injury mitigation. Eur Burn J 4:108–120. https://doi.org/10.3390/EBJ4010011
34. Sharabi M, Mandelberg Y, Benayahu D, Benayahu Y, Azem A, Haj-Ali R (2014) A new class of bio-composite materials of unique collagen fibers. J Mech Behav Biomed Mater 36:71–81. https://doi.org/10.1016/J.JMBBM.2014.04.008
35. Jussila J, Leppäniemi A, Paronen M, Kulomäki E (2005) Ballistic skin simulant. Forensic Sci Int 150:63–71. https://doi.org/10.1016/J.FORSCIINT.2004.06.039
36. Khan A ur R, Huang K, Khalaji MS, Yu F, Xie X, Zhu T et al (2021) Multifunctional bioactive core-shell electrospun membrane capable to terminate inflammatory cycle and promote angiogenesis in diabetic wound. Bioact Mater 6:2783–800. https://doi.org/10.1016/J.BIOACTMAT.2021.01.040
37. Talebian S, Mehrali M, Taebnia N, Pennisi CP, Kadumudi FB, Foroughi J et al (2019) Self-healing hydrogels: the next paradigm shift in tissue engineering? Adv Sci 6. https://doi.org/10.1002/advs.201801664
38. Gupta V, Chanda A (2022) Biomechanics of skin grafts: effect of pattern size, spacing and orientation. Eng Res Express 4:015006. https://doi.org/10.1088/2631-8695/AC48CB
39. Singh G, Chanda A (2023) Development and biomechanical testing of human stomach tissue surrogates. In: Materials horizons: from nature to nanomaterials, Part F1471, pp 113–25. https://doi.org/10.1007/978-981-99-5064-5_7/COVER
40. Wang C, Yang J (2022) Mechanical forces: the missing link between idiopathic pulmonary fibrosis and lung cancer. Eur J Cell Biol 101:151234. https://doi.org/10.1016/J.EJCB.2022.151234
41. Matos HDS, Chu T, Casper BM, Babina MA, Daley MS, Shukla A (2023) Human lung simulants subjected to underwater explosions—an experimental investigation. J Mech Behav Biomed Mater 145:106035. https://doi.org/10.1016/J.JMBBM.2023.106035
42. Balestrini JL, Chaudhry S, Sarrazy V, Koehler A, Hinz B (2012) The mechanical memory of lung myofibroblasts. Integr Biol 4:410–421. https://doi.org/10.1039/C2IB00149G
43. Ionescu C, Lopes A, Copot D, Machado JAT, Bates JHT (2017) The role of fractional calculus in modeling biological phenomena: a review. Commun Nonlinear Sci Numer Simul 51:141–159. https://doi.org/10.1016/J.CNSNS.2017.04.001
44. Liu F, Tschumperlin DJ (2011) Micro-mechanical characterization of lung tissue using atomic force microscopy. JoVE e2911. https://doi.org/10.3791/2911

45. Suki B, Bates JHT (2011) Lung tissue mechanics as an emergent phenomenon. J Appl Physiol 110:1111–1118. https://doi.org/10.1152/JAPPLPHYSIOL.01244.2010/ASSET/IMAGES/LARGE/ZDG0041194870004.JPEG

Chapter 7
Heart Tissue Simulants

The human heart is a vital organ responsible for pumping blood throughout the circulatory system, supplying oxygen and nutrients to the body's tissues, and removing waste products. The heart is composed of several layers of tissue, each with its unique structure and function. Understanding heart tissue is crucial for studying cardiovascular physiology and pathology, and developing treatments for heart-related conditions. Cardiomyocytes are the primary cells in the heart tissue responsible for contraction. These cells have a unique striated appearance and are connected by intercalated discs, which allow for synchronized contraction [1, 2]. Cardiomyocytes are rich in mitochondria to support the high energy demands of the heart. The innermost layer of the heart is called the endocardium. It lines the chambers of the heart and is in direct contact with the blood. It is composed of endothelial cells and connective tissue. The middle layer of the heart is the myocardium, which is primarily composed of cardiac muscle cells (cardiomyocytes). The myocardium is responsible for the pumping action of the heart, and its thickness varies in different parts of the heart [3]. The outer layer of the heart is called the epicardium. It is a thin layer that covers the surface of the heart. The epicardium contains blood vessels, nerves, and connective tissue. The heart is supplied with blood by coronary arteries and drained by coronary veins. The heart is a muscular organ with a complex structure, consisting of four chambers, various valves, and a network of blood vessels. These blood vessels are crucial for providing oxygen and nutrients to the heart tissue. Extracellular Matrix (ECM) provides structural support for heart tissue and plays a role in cell communication [4–6].

A. Chanda and G. Singh, *Soft Tissue Simulants*, Biomedical Materials for Multi-functional Applications, https://doi.org/10.1007/978-981-97-3060-5_7

7.1 Anatomy of the Heart

The heart is located in the thoracic cavity, between the lungs, slightly to the left of the midline. It is enclosed in a protective sac called the pericardium. The human heart is approximately the size of a clenched fist. It has a conical shape with a slightly tilted position, pointing downward, forward, and to the left [7]. The heart is divided into four chambers: two atria (upper chambers) and two ventricles (lower chambers) (see Fig. 7.1). The right atrium receives deoxygenated blood from the body, while the left atrium receives oxygenated blood from the lungs. The right ventricle pumps deoxygenated blood to the lungs, and the left ventricle pumps oxygenated blood to the rest of the body [1]. The heart has four valves that regulate blood flow and prevent backward flow (regurgitation). The atrioventricular (AV) valves are the tricuspid valve on the right side and the bicuspid (mitral) valve on the left side, separating atria from ventricles. The semilunar valves are the pulmonary valve (between the right ventricle and the pulmonary artery) and the aortic valve (between the left ventricle and the aorta). The heart is connected to a network of blood vessels. The pulmonary arteries carry deoxygenated blood from the right ventricle to the lungs, while the pulmonary veins return oxygenated blood to the left atrium. The aorta, the largest artery, carries oxygenated blood from the left ventricle to the rest of the body. The heart has its own blood supply through coronary arteries, which branch off the aorta. Coronary veins collect deoxygenated blood and drain it into the right atrium. The heart is composed of cardiac muscle tissue, which is involuntary and has a unique ability to contract rhythmically and continuously [2, 4].

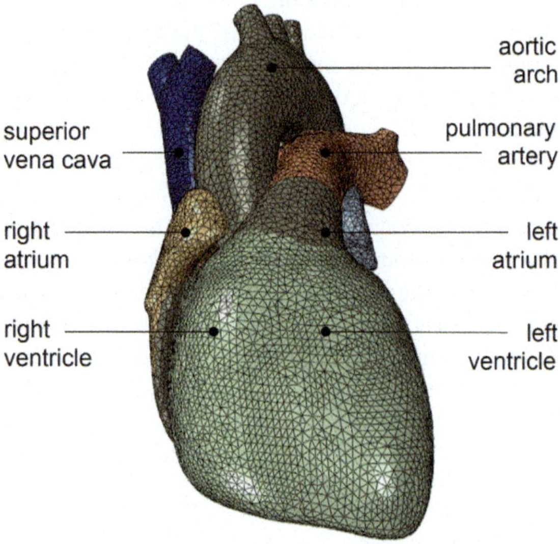

Fig. 7.1 Finite element model of the human heart [8]

7.2 Functions of Heart Tissue

The primary function of heart tissue is to contract rhythmically and continuously to pump blood throughout the circulatory system. The unique property of cardiac muscle is its ability to generate electrical impulses, allowing coordinated contraction and relaxation. The heart tissue, particularly the cardiac muscle cells, contracts in a coordinated manner to pump blood. The atria contract first to push blood into the ventricles, and then the ventricles contract to propel blood into the pulmonary artery and aorta for distribution to the lungs and the rest of the body, respectively. Heart tissue includes components such as heart valves, which play a crucial role in regulating blood flow [7, 9, 10]. Atrioventricular (AV) valves prevent backflow from ventricles to atria during contraction, and semilunar valves prevent backflow from arteries to ventricles during relaxation. The heart pumps deoxygenated blood from the body to the lungs, where it picks up oxygen and releases carbon dioxide. Oxygenated blood is then pumped to the rest of the body, delivering oxygen and essential nutrients to tissues and organs. By pumping blood into the arteries, the heart helps maintain blood pressure, ensuring adequate perfusion of organs and tissues throughout the body [11]. The heart tissue, in coordination with the circulatory system, plays a vital role in maintaining homeostasis by ensuring a balanced and consistent supply of oxygen, nutrients, and other substances to all cells [12].

7.3 Traditional Heart Tissues for Testing

In the field of cardiovascular research and medical testing, various traditional heart tissues are commonly used for experimentation and testing. These tissues provide valuable insights into cardiac function, responses to drugs, and the development of treatments for heart-related conditions. Some of the traditional heart tissues used for testing are from animal models, human cadavers, perfused heart models, etc. Small mammals like mice and rats are commonly used in cardiovascular research due to their availability, cost-effectiveness, and genetic similarities to humans. Isolated hearts or cardiac tissue sections from rodents are often used for testing. The second commonly used animal model for heart tissue testing is porcine heart tissue [13–15]. Pigs (swine) have hearts that are anatomically and physiologically similar to human hearts, making them suitable for certain types of cardiovascular research. Human cadaveric hearts can be used for studying the anatomy and structure of the heart. These tissues are valuable for educational purposes and certain types of research. In addition, thin slices of cardiac tissue, obtained from animal or human hearts, can be used for various experiments [16]. These slices maintain some of the tissue's native structure and function. Perfused heart models are also widely used for testing purposes where hearts can be isolated from animals and perfused with a nutrient-rich solution in a controlled environment. This model allows researchers to study cardiac function, response to drugs, and other parameters under controlled conditions [12, 17].

7.3.1 Challenges with Traditional Heart Tissues

While traditional heart tissues have been valuable for cardiovascular research and testing, there are several challenges associated with their use. These challenges can impact the relevance, reliability, and translatability of experimental findings. Animal hearts, such as those from rodents or pigs, may exhibit physiological and anatomical differences compared to human hearts. These differences can limit the direct translation of findings to human physiology and pathology. Second, human cadaveric tissues are limited in availability, and obtaining them raises ethical concerns. This limitation may restrict the scale and scope of experiments [18–21]. Postmortem changes can occur in tissues, affecting the physiological characteristics. These changes may alter the tissue's responsiveness to experimental conditions and impact the validity of results. There can be significant variability between individuals within a species and even between different batches of animals or tissue samples. This variability can introduce confounding factors and affect the reproducibility of experiments. The conditions in which traditional heart tissues are tested (e.g., in vitro cultures or perfusion systems) may not fully mimic the in vivo environment [22–24]. Dynamic factors such as blood flow, hormonal regulation, and neural input are challenging to replicate accurately. Standardizing experimental conditions across different laboratories can be difficult. Variations in techniques, reagents, and equipment can lead to discrepancies in results and hinder the comparability of studies [12].

Many traditional heart tissue models may not fully capture the complexity of cardiovascular diseases, limiting their utility for studying disease mechanisms and testing therapeutic interventions. Researchers are continually working to address these challenges by developing more sophisticated models, such as organoids, microphysiological systems, and humanized animal models [25, 26]. Additionally, advancements in tissue engineering and stem cell technologies aim to create more physiologically relevant and scalable platforms for cardiovascular research. However, the best-suited alternative for traditional heart tissue is the development of artificial tissue simulants which can mimic the properties of real tissue and have no biosafety or handling issues.

7.4 Heart Tissue Simulants

Heart tissue simulants are materials designed to mimic the mechanical, electrical, and biological properties of real heart tissue. These simulants play a crucial role in various medical and engineering applications, including the development and testing of medical devices, the study of cardiac mechanics, and the training of healthcare professionals. Heart tissue simulants should replicate the elastic nature of real heart tissue. This involves mimicking the compliance and stiffness that the heart exhibits during contraction and relaxation. The heart muscle fibers have a specific orientation, and simulants may aim to replicate this anisotropic behavior to accurately represent the mechanical characteristics of the myocardium [15, 27].

Heart tissue simulants should possess electrical conductivity similar to that of real cardiac tissue. This is crucial for simulating the electrical conduction that occurs during the heart's rhythmic contractions. Simulants may be designed to reproduce the electrical excitability, action potential, and propagation characteristics of heart tissue [28, 29]. In some cases, simulants may need to be biocompatible to allow interactions with biological tissues, such as in the development and testing of implantable devices. For certain applications, simulants may incorporate optical properties to enable imaging techniques such as fluorescence or light scattering studies [12].

7.5 Types of Heart Tissue Simulants

Hydrogels are water-absorbing polymers that can mimic the mechanical properties of soft tissues, including heart tissue. They can be engineered to mimic the elasticity and structure of heart tissue. Synthetic elastomers, another type of heart tissue simulants may be used to replicate the elastic properties of cardiac tissue. Silicones, polyurethanes, and other elastomers are often used to fabricate flexible and stretchable materials for cardiovascular simulations [13–15]. To mimic the electrical conductivity of cardiac tissue, conductive polymers may be integrated into simulants. These materials can simulate the electrical behavior of the heart. Simulants may be developed by combining different materials to achieve a balance of mechanical and electrical properties [25]. For example, a composite of hydrogel and conductive particles may be used. Some researchers use animal tissues, such as latex or agar gel models, to simulate certain aspects of heart tissue. While not as accurate as real tissue, these surrogates can provide a simplified representation for specific experiments [12].

Advances in medical imaging allow for the creation of patient-specific models using 3D printing or virtual simulations. These models enable personalized medicine and help in understanding individual variations in cardiac anatomy and function. 3D bioprinting technology allows researchers to create structures that mimic realistic heart tissue [30, 31]. It's important to note that each simulant has its limitations, and researchers often choose the most appropriate model based on the specific aspects of heart tissue they aim to study or replicate. Combining different simulants and technologies allows for a more comprehensive understanding of cardiac physiology and pathology.

7.6 Applications of Heart Tissue Simulants

Heart tissue simulants are valuable for testing and optimizing the performance of medical devices such as pacemakers, defibrillators, and cardiac catheters. Simulants are used in medical training scenarios, allowing healthcare professionals to practice procedures and interventions in a controlled environment. Simulants aid researchers in studying cardiac mechanics, electrophysiology, and the development of new therapeutic interventions [12].

7.6.1 Medical Device Testing

Simulants allow researchers to test and evaluate the performance of cardiovascular devices such as stents, pacemakers, and heart valves in a controlled environment before human trials. Heart tissue simulants allow researchers and engineers to test prototypes of cardiovascular devices in a controlled laboratory setting. This is crucial in the early stages of device development to identify design flaws, assess functionality, and make necessary improvements before moving to more advanced testing stages. Simulants are used to assess the mechanical properties of cardiovascular devices [22, 32]. For example, if a device interacts with the heart muscle, simulants can replicate the mechanical behavior of cardiac tissue to evaluate how the device responds to different forces, pressures, and deformations. Heart tissue simulants are employed to mimic the blood flow and pressure conditions within the cardiovascular system. This is particularly important for testing devices such as stents, valves, and vascular grafts to ensure that they perform optimally under realistic physiological conditions. Simulants are used to simulate long-term usage scenarios to assess the durability of cardiovascular devices [33, 34]. This involves subjecting the devices to cyclic loading, mimicking the constant beating of the heart, and evaluating how well they withstand wear and tear over time. Using heart tissue simulants can reduce the reliance on animal testing, providing a more cost-effective and ethical alternative. This is particularly beneficial in the early stages of device development and optimization [12].

In summary, heart tissue simulants offer a versatile and valuable tool for testing cardiovascular devices. They facilitate rigorous evaluation under controlled conditions, ensuring that devices meet safety and performance standards before progressing to clinical trials and eventual patient use. This approach contributes to the development of safer and more effective medical devices for the treatment of cardiovascular conditions.

7.6.2 Biomechanical Studies

Heart tissue simulants are employed to replicate the elasticity and compliance of actual heart tissue. The aim is to characterize the mechanical properties of the simulants to closely match those of the heart, allowing for accurate representation of tissue behavior during different phases of the cardiac cycle. Biomechanical studies often involve testing different materials for their suitability in mimicking heart tissue [22]. Simulants made from synthetic materials, such as elastomers or hydrogels, are chosen and validated based on their ability to replicate the mechanical responses of real heart tissue. Heart tissue simulants can be designed to simulate the contractile and relaxative properties of the heart muscle. This allows researchers to study how the tissue responds to changes in pressure and volume during the cardiac cycle, aiding in the understanding of normal and pathological cardiac function [28, 35].

Biomechanical studies involve subjecting heart tissue simulants to controlled mechanical loads to analyze stress and strain distributions. This helps researchers understand how the tissue deforms under different conditions, providing insights into potential areas of stress concentration and deformation. Further, heart tissue simulants are used as training models for surgeons and medical professionals. Simulating the mechanical properties of real heart tissue allows trainees to practice surgical procedures, such as suturing and tissue manipulation, in a realistic environment without the need for cadavers or live animals. Biomechanical studies involving heart tissue simulants can incorporate dynamic flow simulations to replicate blood flow conditions [12, 36]. This is particularly relevant for assessing the impact of different cardiovascular conditions on the mechanical behavior of the heart and its response to varying hemodynamic forces. The precisely developed heart tissue simulants can be modified to simulate the mechanical changes associated with specific cardiac pathologies, such as hypertrophy or fibrosis. This allows researchers to study how these conditions affect the biomechanics of the heart and explore potential interventions [12].

In summary, heart tissue simulants play a pivotal role in biomechanical studies, providing researchers with a controlled and reproducible environment to investigate the mechanical aspects of the heart. These studies contribute to our understanding of cardiac function, aid in the development of new therapies, and enhance the training of medical professionals involved in cardiovascular care.

7.6.3 Surgical Training

The application of heart tissue simulants in surgical training is a valuable and impactful use of these materials. Simulants offer a realistic and controlled environment for training surgeons and other medical professionals, allowing them to practice and refine their skills before working with actual patients. Heart tissue simulants are designed to closely mimic the anatomical and textural properties of real heart tissue.

This includes replicating the feel, elasticity, and response to surgical instruments, providing a realistic training experience for surgeons [37, 38]. Simulants allow surgeons to practice various cardiac surgical procedures, such as suturing, tissue manipulation, and knot tying. This is especially important for mastering delicate and precise techniques required in cardiac surgery. With the rise of minimally invasive techniques in cardiac surgery, heart tissue simulants are used to train surgeons in performing procedures using endoscopic or robotic tools. Simulants provide a platform to develop proficiency in these advanced techniques without the need for live tissue or cadavers. Heart tissue simulants can be customized to simulate specific pathological conditions, such as coronary artery disease or valve disorders. This allows surgeons to practice procedures tailored to address these conditions and develop strategies for managing complications [21, 39]. Simulants are used to train surgeons in the precise placement of devices like stents and prosthetic valves. This ensures that surgeons are skilled in navigating the cardiovascular system and deploying devices with accuracy. Surgical training with heart tissue simulants significantly reduces the risks associated with practicing on live patients. This ethical and risk-free learning environment is particularly important for complex and high-risk procedures [22].

In summary, the application of heart tissue simulants in surgical training provides an effective and ethical means for surgeons to develop and enhance their skills. The realistic simulation of cardiac procedures helps improve patient outcomes by ensuring that surgeons are well-prepared and proficient in performing intricate cardiovascular surgeries.

7.6.4 Educational Purposes

The application of heart tissue simulants in educational settings is multifaceted, encompassing various aspects of medical education. Heart tissue simulants serve as valuable tools for teaching and learning about cardiac anatomy, physiology, and medical procedures. Heart tissue simulants are used to create anatomically accurate models of the heart, allowing students to study the structure and arrangement of cardiac chambers, valves, and vessels. This hands-on approach enhances understanding and retention of complex anatomical details [18, 19]. In medical education, students, especially those pursuing surgical specialties, use heart tissue simulants to practice basic surgical skills such as suturing and knot tying. These foundational skills are crucial for various surgical procedures, including those related to the heart. Simulants can be altered to simulate specific cardiac pathologies, allowing students to study the impact of diseases on heart function and structure [20, 40]. This enhances student comprehension of cardiovascular diseases and their clinical manifestations. Heart tissue simulants can mimic the feel and sounds of a real heart, enabling students to practice palpation and auscultation skills. This is essential for developing the ability to detect abnormalities in heart sounds and rhythms. Simulants are used to practice echocardiography techniques. Students can learn how to use ultrasound probes, interpret images, and identify different cardiac structures, contributing to their proficiency in diagnostic imaging [12].

In summary, the application of heart tissue simulants in educational settings contributes to the comprehensive training of healthcare professionals. These simulants enhance hands-on learning, improve procedural skills, and provide a safe and ethical environment for students to gain valuable experience in cardiovascular medicine.

7.7 Challenges with Heart Tissue Simulants

Achieving a comprehensive replication of the complex mechanical and electrical properties of the heart is challenging. Incorporating the dynamic and evolving nature of real heart tissue, especially in disease conditions, can be difficult. Heart tissue simulants contribute significantly to advancements in medical technology, research, and education by providing a controlled and replicable platform for experimentation and testing. Their development continues to be an active area of research, aiming to create more accurate representations of the intricate properties of the human heart [9, 26]. While heart tissue simulants offer valuable benefits for various applications, there are challenges associated with their use. These challenges can impact the accuracy of simulation and the extent to which simulants replicate the complexity of real heart tissues.

Replicating the intricate mechanical and biological properties of real heart tissue is challenging. The heart is a dynamic organ with unique properties such as anisotropy, nonlinearity, and time-dependent behavior. Simulants may struggle to fully capture these complexities. Heart tissue simulants may lack the cellular and microstructural features found in real cardiac tissues [11, 25]. Mimicking the arrangement of cardiomyocytes, blood vessels, and extracellular matrix with high fidelity is difficult, and the absence of these features can limit the realism of the simulation. Simulants may not accurately represent the biocompatibility of real tissues or the dynamic interactions between medical devices and the surrounding biological environment. Understanding how the simulated tissue responds to interventions and foreign bodies is crucial for reliable testing. Some simulants may degrade or change their mechanical properties over time, affecting the reliability of long-term studies. Ensuring the stability of simulants throughout the experiment, especially those involving cyclic loading, is a significant challenge [15, 17]. Expense and Availability: High-quality heart tissue simulants can be expensive to produce, and their availability may be limited. This can be a barrier for researchers and educators, especially those with budget constraints or limited access to specialized materials. There is a lack of standardized protocols for the preparation and characterization of heart tissue simulants. This can lead to variations in material properties between different laboratories, affecting the reproducibility and comparability of research outcomes [13, 14]. While heart tissue simulants provide valuable training opportunities, they may not fully replicate the complexity of live tissue. This limitation can impact the realism of surgical training scenarios, especially for advanced or complex procedures. In certain educational or research settings, there may be ethical considerations regarding the use of live animal or human tissues. While stimulants offer an ethical alternative, they may not completely replace the need for live tissue in some scenarios [12].

Addressing these challenges requires ongoing research and development efforts in the field of biomaterials, tissue engineering, and simulation technology. Collaboration between researchers, clinicians, and educators is essential to overcome these limitations and advance the effectiveness and realism of heart tissue simulants in various applications.

References

1. Stradins P (2004) Comparison of biomechanical and structural properties between human aortic and pulmonary valve. Eur J Cardio-Thoracic Surg 26:634–639. https://doi.org/10.1016/j.ejcts.2004.05.043
2. Holzapfel GA, Sommer G, Gasser CT, Regitnig P (2005) Determination of layer-specific mechanical properties of human coronary arteries with nonatherosclerotic intimal thickening and related constitutive modeling. Am J Physiol—Hear Circ Physiol 289:2048–2058. https://doi.org/10.1152/ajpheart.00934.2004
3. Sack I, Rump J, Elgeti T, Samani A, Braun J (2009) MR elastography of the human heart: noninvasive assessment of myocardial elasticity changes by shear wave amplitude variations. Magn Reson Med 61:668–677. https://doi.org/10.1002/mrm.21878
4. Canham PB, Finlay HM, Dixon JG, Boughner DR, Chen A (1989) Measurements from light and polarised light microscopy of human coronary arteries fixed at distending pressure. Cardiovasc Res 23:973–982. https://doi.org/10.1093/cvr/23.11.973
5. Richardson PD, Keeny SM (1989) Anisotropy of human coronary artery intima. In: Bioengineering proceedings of northeast conference, IEEE, pp 205–206. https://doi.org/10.1109/nebc.1989.36772
6. Oveissi F, Naficy S, Lee A, Winlaw DS, Dehghani F (2020) Materials and manufacturing perspectives in engineering heart valves: a review. Mater Today Bio 5:100038. https://doi.org/10.1016/j.mtbio.2019.100038
7. Nash MP, Hunter PJ (2000) Computational mechanics of the heart. From tissue structure to ventricular function. J Elast 61:113–141. https://doi.org/10.1023/A:1011084330767/METRICS
8. Baillargeon B, Rebelo N, Fox DD, Taylor RL, Kuhl E (2014) The living heart project: a robust and integrative simulator for human heart function. Eur J Mech—A/Solids 48:38–47. https://doi.org/10.1016/J.EUROMECHSOL.2014.04.001
9. Mendelson K, Schoen FJ (2006) Heart valve tissue engineering: concepts, approaches, progress, and challenges. Ann Biomed Eng 34:1799–1819. https://doi.org/10.1007/S10439-006-9163-Z/FIGURES/5
10. Courtney T, Sacks MS, Stankus J, Guan J, Wagner WR (2006) Design and analysis of tissue engineering scaffolds that mimic soft tissue mechanical anisotropy. Biomaterials 27:3631–3638. https://doi.org/10.1016/J.BIOMATERIALS.2006.02.024
11. Kapnisi M, Mansfield C, Marijon C, Guex AG, Perbellini F, Bardi I et al (2018) Auxetic cardiac patches with tunable mechanical and conductive properties toward treating myocardial infarction. Adv Funct Mater 28:1800618. https://doi.org/10.1002/ADFM.201800618
12. Singh G, Chanda A (2021) Mechanical properties of whole-body soft human tissues: a review. Biomed Mater 16:062004. https://doi.org/10.1088/1748-605X/AC2B7A

13. Kalcioglu ZI, Mrozek RA, Mahmoodian R, VanLandingham MR, Lenhart JL, Van Vliet KJ (2013) Tunable mechanical behavior of synthetic organogels as biofidelic tissue simulants. J Biomech 46:1583–1591. https://doi.org/10.1016/J.JBIOMECH.2013.03.011
14. Clayton RH, Bernus O, Cherry EM, Dierckx H, Fenton FH, Mirabella L et al (2011) Models of cardiac tissue electrophysiology: progress, challenges and open questions. Prog Biophys Mol Biol 104:22–48. https://doi.org/10.1016/J.PBIOMOLBIO.2010.05.008
15. Laflamme MA, Murry CE (2011) Heart regeneration. Nat 2011 4737347 2011;473:326–335. https://doi.org/10.1038/nature10147
16. Nash MP, Panfilov AV (2004) Electromechanical model of excitable tissue to study reentrant cardiac arrhythmias. Prog Biophys Mol Biol 85:501–522. https://doi.org/10.1016/J.PBIOMOLBIO.2004.01.016
17. Driessen NJB, Boerboom RA, Huyghe JM, Bouten CVC, Baaijens FPT (2003) Computational analyses of mechanically induced collagen fiber remodeling in the aortic heart valve. J Biomech Eng 125:549–557. https://doi.org/10.1115/1.1590361
18. Chanda A, Singh G (2023) Applications, challenges, and future opportunities. Mater Horiz Nat Nanomater 85–92. https://doi.org/10.1007/978-981-99-2225-3_8/COVER.
19. Chanda A, Unnikrishnan V, Lackey K, Robbins J (2020) Biofidelic conductive soft tissue surrogates. Int J Polym Mater Polym Biomater 69:127–135. https://doi.org/10.1080/00914037.2018.1552856
20. Singh G, Chanda A (2023) Biofidelic tongue and tonsils tissue surrogates. Mater Horiz Nat Nanomater Part F1471:159–70. https://doi.org/10.1007/978-981-99-5064-5_10/COVER
21. Singh G, Chanda A (2023) Development and biomechanical testing of human stomach tissue surrogates. Mater Horiz Nat Nanomater, Part F1471:113–25. https://doi.org/10.1007/978-981-99-5064-5_7/COVER
22. Singh G, Chanda A (2023) Development and mechanical characterization of artificial surrogates for brain tissues. Biomed Eng Adv 5:100084. https://doi.org/10.1016/J.BEA.2023.100084
23. Gupta V, Singh G, Chanda A (2022) Development and testing of skin grafts models with varying slit orientations. Mater Today Proc 62:3462–3467. https://doi.org/10.1016/J.MATPR.2022.04.282
24. Makode S, Singh G, Chanda A (2021) Development of novel anisotropic skin simulants. Phys Scr 96:125019. https://doi.org/10.1088/1402-4896/AC2EFD
25. Knöll R, Hoshijima M, Chien K (2003) Cardiac mechanotransduction and implications for heart disease. J Mol Med 81:750–756. https://doi.org/10.1007/S00109-003-0488-X/TABLES/1
26. Garrett AD (2013) Dabigatran vs. Warfarin in patients with mechanical heart valves. Drug Top 369:1206–1220. https://doi.org/10.1056/NEJMOA1300615/SUPPL_FILE/NEJMOA1300615_DISCLOSURES.PDF
27. Griendling KK, Gerthoffer WT (2007) Mechanisms of vascular smooth muscle cell migration. Circ Res 100:607–621. https://doi.org/10.1161/01.RES.0000258492.96097.47
28. Chanda A, Singh G (2023) Introduction to human tissues. Mater Horiz Nat Nanomater 1–12. https://doi.org/10.1007/978-981-99-2225-3_1/COVER
29. Gupta V, Singh G, Gupta S, Chanda A (2023) Expansion potential of auxetic prosthetic skin grafts: a review. Eng Res Express 5:022003. https://doi.org/10.1088/2631-8695/ACCFE5
30. Singh G, Gupta V, Chanda A (2022) Artificial skin with varying biomechanical properties. Mater Today Proc 62:3162–3166. https://doi.org/10.1016/J.MATPR.2022.03.433
31. Chanda A, Callaway C (2018) Tissue anisotropy modeling using soft composite materials. Appl Bionics Biomech 2018. https://doi.org/10.1155/2018/4838157
32. Chanda A, Singh G (2023) Muscles and connective tissues. Mater Horiz Nat Nanomater 25–32. https://doi.org/10.1007/978-981-99-2225-3_3/COVER
33. Liu Z, Khalil RA (2018) Evolving mechanisms of vascular smooth muscle contraction highlight key targets in vascular disease. Biochem Pharmacol 153:91–122. https://doi.org/10.1016/J.BCP.2018.02.012
34. Hunter PJ, McCulloch AD, Ter Keurs HEDJ (1998) Modelling the mechanical properties of cardiac muscle. Prog Biophys Mol Biol 69:289–331. https://doi.org/10.1016/S0079-6107(98)00013-3

35. Chanda A, Singh G (2023) Hyperelastic models for anisotropic tissue characterization. Mater Horiz Nat Nanomater 73–83. https://doi.org/10.1007/978-981-99-2225-3_7/COVER
36. Chanda A, McClain S (2019) Mechanical modeling of healthy and diseased calcaneal fat pad surrogates. Biomimetics 4(1):1. https://doi.org/10.3390/BIOMIMETICS4010001
37. Chanda A (2018) Biomechanical modeling of human skin tissue surrogates. Biomimetics 3(3):18. https://doi.org/10.3390/BIOMIMETICS3030018
38. Gupta V, Chanda A (2022) Biomechanics of skin grafts: effect of pattern size, spacing and orientation. Eng Res Express 4:015006. https://doi.org/10.1088/2631-8695/AC48CB
39. Singh G, Chanda A (2023) Development and biomechanical testing of artificial surrogates for vaginal tissue. Adv Mater Process Technol. https://doi.org/10.1080/2374068X.2023.2198837
40. Singh G, Chanda A (2023) Biofidelic gallbladder tissue surrogates. Adv Mater Process Technol. https://doi.org/10.1080/2374068X.2023.2198835

Chapter 8
Breast Tissue Simulants

Breast tissue is a complex and dynamic structure that consists of a combination of glandular, connective, and adipose (fat) tissues, and its primary function is to produce and secrete milk for breastfeeding [1, 2]. It undergoes changes throughout a woman's life, particularly during puberty, menstrual cycle, pregnancy, and lactation. The primary function of breast tissue is to produce and secrete milk to nourish an infant during breastfeeding.

8.1 Anatomy of the Breast Tissue

The primary component of the breast tissue is glandular or mammary glands. These are the functional components responsible for milk production. They are arranged in lobules and connected by ducts that transport milk to the nipple. Fibrous bands are the connective tissue of the breast and provide structural support to the breast tissue, helping to maintain its shape. Adipose tissue surrounds and cushions the glandular and connective tissues. The amount of adipose tissue determines breast size and varies among individuals [3–7]. The breast is supplied with blood by arteries, and blood is drained through veins. Lymphatic vessels in the breast play a crucial role in draining excess fluid and filtering waste products. Lymph nodes, particularly in the axillary region, are important in breast cancer staging. The raised structure at the center of the breast through which milk is expelled during breastfeeding is the nipple and the pigmented area surrounding the nipple is known as areola [8]. Figure 8.1 shows the structure of the breast tissue and its changes during lactation [9].

A. Chanda and G. Singh, *Soft Tissue Simulants*, Biomedical Materials for Multi-functional Applications, https://doi.org/10.1007/978-981-97-3060-5_8

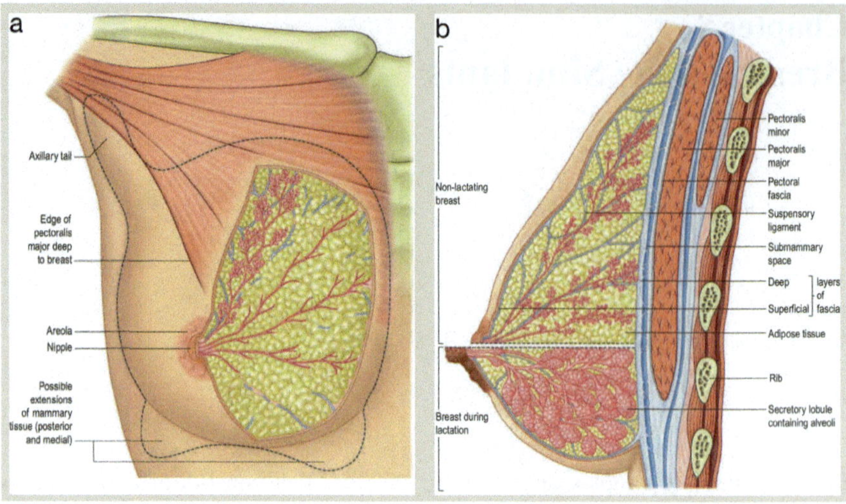

Fig. 8.1 The structure of the breast (**a**) and changes in breast structure during lactation (**b**) [9]

8.1.1 Functions of the Breast Tissue

Glandular tissue is responsible for producing milk during lactation. Hormonal signals, particularly from the pituitary gland, regulate the synthesis and release of milk. Breast milk provides essential nutrients, antibodies, and immune factors for newborns, contributing to their optimal growth and development. Breast tissue is responsive to hormonal changes, particularly estrogen and progesterone, which influence the menstrual cycle, pregnancy, and lactation [10, 11].

8.2 Need of Breast Tissue for Testing and Research Purposes

Breast tissue is essential for testing and research purposes in various fields of medicine and science. First, breast tissue is crucial for studying the biology and pathology of breast cancer. Researchers use breast tissue samples to investigate the genetic, molecular, and cellular changes associated with different types of breast cancer. This research is fundamental for developing targeted therapies and improving treatment outcomes. Second, breast tissue is used to test the effectiveness and safety of new drugs and treatments for breast cancer. This helps researchers identify potential therapeutic agents and understand how different drugs interact with breast tissue at the cellular and molecular levels [12–14]. Third, studying normal breast tissue is important for understanding its physiology and function. This knowledge is essential for identifying abnormalities, such as those related to breast development, lactation, and hormonal regulation. Fourth, breast tissue is used in the development and testing of diagnostic tools, such as imaging techniques (e.g., mammography, MRI) and biopsy procedures. Improving diagnostic accuracy is crucial for early detection of breast cancer and initiating timely treatment. Fifth, researchers use breast tissue in the field of tissue engineering to develop synthetic or bioengineered tissues for reconstructive surgeries. This is particularly relevant for breast reconstruction after mastectomy. Sixth, biomarkers derived from breast tissue can be used for early detection, prognosis, and monitoring treatment response in breast cancer patients. These biomarkers help tailor treatment plans to individual patients [15–17]. Seventh, medical professionals, including pathologists and surgeons, use breast tissue samples for educational and training purposes. This allows them to practice techniques such as biopsy interpretation and surgical procedures before performing them on patients. Eighth, breast tissue research contributes to our understanding of the epidemiology of breast cancer, including risk factors, prevalence, and distribution. This information is valuable for public health initiatives and preventive strategies [13, 15].

In summary, breast tissue is indispensable for advancing our understanding of breast health, diagnosing and treating breast cancer, and developing innovative medical interventions. It plays a critical role in multidisciplinary research efforts aimed at improving patient outcomes and addressing various aspects of breast-related health.

8.3 Traditional Breast Tissues for Testing

When it comes to testing and research related to breast tissues, various traditional models have been used before the advent of advanced tissue simulants and three-dimensional cell cultures. These traditional models have played important roles in understanding breast physiology, pathology, and the development of diseases. Cultured cell lines derived from breast tissues have been widely used for in vitro experiments. These are populations of cells that have been adapted to grow continuously in a laboratory setting. Cell lines are utilized to study cellular behavior, responses to stimuli, and to model specific breast cancer subtypes. MCF-7 and MDA-MB-231 are examples of well-known breast cancer cell lines [18–20]. Animal models, such as mice and rats, have been employed for in vivo experiments to study breast development, cancer progression, and the effects of potential treatments. These models are used to investigate physiological processes, test the efficacy of drugs, and understand the role of specific genes in breast-related conditions. In addition to these two models, xenograft models are also widely used. Xenograft models involve the transplantation of human breast tissue or breast cancer cells into immunocompromised mice. These models are used to study tumor growth, invasiveness, and responses to therapies in an in vivo setting [21].

While these traditional models have provided valuable insights into breast biology and pathology, they have limitations, such as a lack of dynamic physiological conditions and challenges in replicating the complexity of the human breast microenvironment. The development of more advanced tissue simulants and three-dimensional cell culture models has addressed some of these limitations, offering more sophisticated tools for breast tissue testing and research.

8.3.1 Challenges with Traditional Breast Tissues

Traditional approaches to studying breast tissues, such as the use of cell lines, animal models, histological sections, and other methods, have been foundational in breast cancer research and understanding breast physiology. However, these approaches come with various challenges and limitations, and advancements in technology and methodology aim to address some of these issues. Traditional models often lack reproducibility due to inherent biological variability, especially when working with animal models or primary tissues from different donors [17, 22]. This can affect the consistency and reliability of experimental results. Cell lines and monolayer cultures do not fully capture the three-dimensional (3D) complexity of breast tissue. This can limit the translatability of findings, as the microenvironment and cell–cell interactions in 3D are crucial for understanding tissue behavior. Findings from animal models may not always directly translate to human biology due to species differences. Variations in physiology, metabolism, and immune response can impact the relevance of animal model data to human conditions. The use of animal models raises ethical concerns, particularly regarding the welfare of the animals [13, 23]. Ethical considerations also arise in the use of human tissues, prompting the need for informed consent and privacy protection. Human breast tissues, especially healthy tissues, may be scarce and difficult to obtain in large quantities for research purposes. This limitation can constrain the scope and scale of experiments.

To address these challenges, researchers are increasingly turning to advanced techniques, such as three-dimensional organoid cultures, tissue engineering approaches, and making breast tissue simulants that aim to better replicate the physiological and pathological conditions of breast tissues [24]. While traditional models have been essential for foundational research, ongoing advancements aim to provide more accurate and clinically relevant platforms for studying breast tissues and diseases.

8.4 Need of Breast Tissue Simulants

Breast tissue simulants are synthetic or engineered materials designed to mimic the characteristics of real breast tissue. These simulants are valuable tools in various applications, including medical research, device testing, and training [24]. Here are some reasons why breast tissue simulants are important and extensively used in recent scenarios.

Medical Device Testing: Simulants allow for the testing and evaluation of medical devices designed for breast applications, such as imaging devices (e.g., mammography, ultrasound), biopsy tools, and breast implants. They provide a realistic medium for practicing image interpretation and diagnostic procedures [24].

Training and Education: Breast tissue simulants are used in medical training programs to provide students, healthcare professionals, and surgeons with a realistic and safe environment for practicing procedures. This includes techniques related to breast examinations, biopsies, and surgical interventions [24].

Quality Control in Imaging: Simulants are employed in quality control processes for breast imaging equipment. By using breast tissue simulants with known characteristics, medical professionals can assess and monitor the performance of imaging devices, ensuring accurate and reliable results [24].

Research on Breast Imaging Techniques: Researchers use breast tissue simulants to evaluate and improve breast imaging techniques. This includes developing new imaging modalities, optimizing existing technologies, and studying the interactions between different tissues and imaging devices [24].

Development of Breast Cancer Models: Breast tissue simulants play a role in developing realistic models for studying breast cancer. These models help researchers investigate the progression of the disease, test potential treatments, and understand the behavior of cancerous cells within a simulated breast tissue environment [24].

Cosmetic Surgery Training: In the field of cosmetic and reconstructive surgery, breast tissue simulants are used for training purposes. Surgeons can practice breast augmentation, reduction, and reconstruction procedures, enhancing their skills and precision before performing surgeries on live patients [24].

Testing Cosmetic Products: Breast tissue simulants are employed in the cosmetic industry to test the safety and efficacy of products such as creams, lotions, and treatments designed for breast enhancement or skin care [24].

Biomechanics Studies: Simulants are used in biomechanics studies to understand the mechanical properties of breast tissue. This knowledge is crucial for designing supportive garments, such as bras, and for developing technologies related to breast health and comfort [24].

Development of Prosthetics and Wearable Devices: Breast tissue simulants contribute to the development and testing of prosthetics, breast forms, and wearable devices designed to support individuals who have undergone mastectomy or breast-related surgeries [24].

Radiation Therapy Planning: Simulants are utilized in radiation therapy planning for breast cancer treatment. They help simulate the response of breast tissue to radiation, allowing healthcare professionals to optimize treatment plans and minimize damage to healthy tissues [24].

Breast tissue simulants serve as important tools in various disciplines, offering realistic alternatives for research, testing, and training purposes without the ethical concerns and limitations associated with using live tissues. Their use contributes to advancements in breast health, medical technology, and patient care.

8.5 Composition and Materials for Breast Tissue Simulants

Various types of breast tissue simulants have been developed to meet different needs in medical education, research, and testing. These simulants aim to replicate the mechanical, structural, and radiological properties of natural breast tissue. Some of these are silicone and elastomer-based simulants, hydrogel-based simulants, polyurethane foam simulants, phantom models with inclusions, 3D-printed simulants, etc. Elastomers and silicones are commonly used for breast tissue simulants due to their flexibility, elasticity, and durability. They can mimic the softness and compressibility of natural breast tissue [23]. These materials offer good repeatability, allowing consistent testing and simulations. They can be customized to achieve specific mechanical properties. After silicone-based simulants, hydrogels are the second most commonly used material to mimic breast tissue. Hydrogels are water-containing polymers with properties similar to natural tissues. They can replicate the hydration and texture of breast tissue. Hydrogels provide a more realistic feel and response to pressure, making them suitable for applications where tissue hydration is important [15, 16].

Polyurethane foams are lightweight and can be engineered to mimic the density and compressibility of breast tissue. These foams are often used in breast phantoms for mammography and ultrasound training due to their ability to simulate the radiographic characteristics of breast tissue. Thermoplastic polymers, such as thermoplastic elastomers (TPE), can be molded to replicate the mechanical properties of breast tissue. These materials are versatile and can be customized for specific applications. They are often used in the development of breast models for clinical training. Fabric-based simulants incorporate textile materials to mimic the layered structure of breast tissue [25–27]. These simulants can offer a more realistic representation of tissue layers and can be used in applications like biopsy training or surgical simulations. Phantom models with inclusions combine different materials to simulate various breast tissue components, such as glandular tissue, adipose tissue, and tumors. These tissue models are suitable for testing medical imaging techniques, including ultrasound, mammography, and MRI. Composite-type breast tissue simulants utilize a combination of materials, such as silicone, rubber, and gel, to achieve a more comprehensive representation of breast tissue properties. 3D-printed simulants use 3D printing technology to create customizable breast tissue simulants with specific anatomical features [25, 28, 29]. These simulants can be tailored for surgical training, education, and research purposes, offering a high level of detail (see

Fig. 8.2). Designed with biodegradable materials to address environmental concerns associated with the disposal of synthetic materials, these are suitable for ethical and environmentally conscious educational and research purposes [16, 23, 24].

Each type of breast tissue simulant has its advantages and applications, and the choice depends on the specific requirements of the intended use. Researchers often select simulants based on factors such as realism, durability, customizability, and ethical considerations. Advances in materials science and technology continue to contribute to the development of more sophisticated and versatile breast tissue simulants [28].

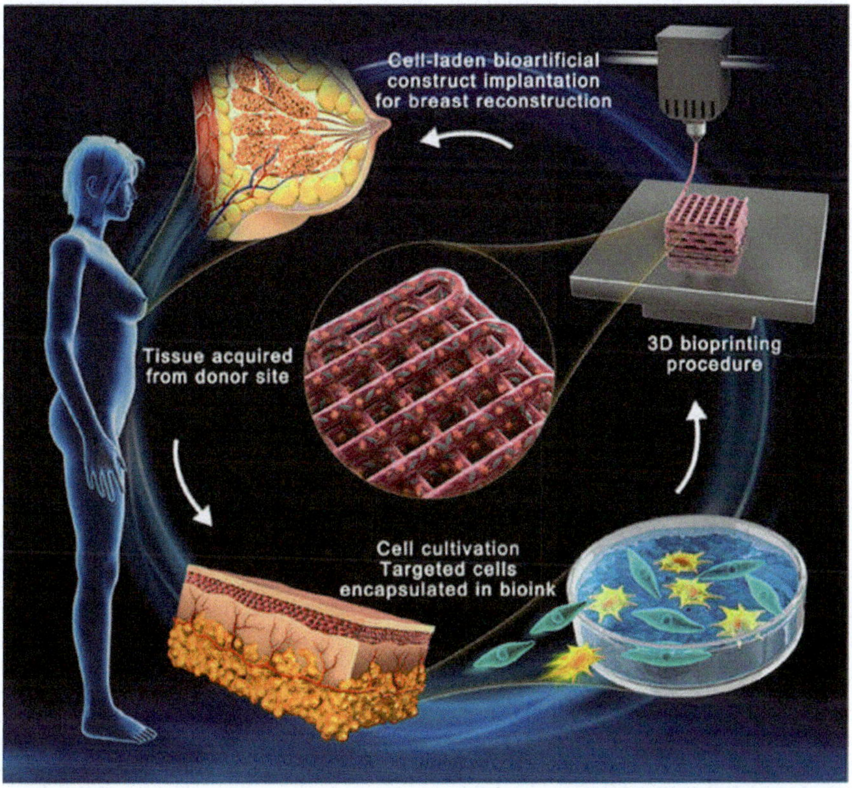

Fig. 8.2 Reconstruction of breast with personalized three-dimensional (3D) bio-printed adipose tissue [30]

8.6 Challenges and Considerations

Achieving a perfect match to the complex biological properties of natural breast tissue is challenging, and simulants may not fully replicate certain characteristics. Some simulants may exhibit changes in mechanical properties over time, affecting their stability and reliability for long-term studies. Standardizing the properties of breast tissue simulants across different laboratories is a challenge. Validation against real tissues or established benchmarks is crucial for ensuring accuracy. While simulants address ethical concerns associated with live tissue use, there are ethical considerations in the production and use of synthetic materials, especially if they involve animal testing or non-biodegradable components. In conclusion, breast tissue simulants are valuable tools in medical education, research, and device testing [20, 31]. Their continued development and refinement contribute to improved training methodologies and advancements in breast-related healthcare.

References

1. Ramião NG, Martins PS, Rynkevic R, Fernandes AA, Barroso M, Santos DC (2016) Biomechanical properties of breast tissue, a state-of-the-art review. Biomech Model Mechanobiol 15:1307–1323. https://doi.org/10.1007/s10237-016-0763-8
2. Krouskop TA, Wheeler TM, Kallel F, Garra BS, Hall T (1998) Elastic moduli of breast and prostate tissues under compression. Ultrason Imaging 20:260–274. https://doi.org/10.1177/016173469802000403
3. Gold RH, Montgomery CK, Minagi H, Annes GP (1971) The significance of mammary skin thickening in disorders other than primary carcinoma: a roentgenologic-pathologic correlation. Am J Roentgenol Radium Ther Nucl Med 112:613–621. https://doi.org/10.2214/ajr.112.3.613
4. Willson SA, Adam EJ, Tucker AK (1982) Patterns of breast skin thickness in normal mammograms. Clin Radiol 33:691–693. https://doi.org/10.1016/S0009-9260(82)80407-8
5. Huang SY, Boone JM, Yang K, Kwan ALC, Packard NJ (2008) The effect of skin thickness determined using breast CT on mammographic dosimetry. Med Phys 35:1199–1206. https://doi.org/10.1118/1.2841938
6. Ulger H, Erdogan N, Kumanlioglu S, Unur E (2003) Effect of age, breast size, menopausal and hormonal status on mammographic skin thickness. Ski Res Technol 9:284–289. https://doi.org/10.1034/j.1600-0846.2003.00027.x
7. Sutradhar A, Miller MJ (2013) In vivo measurement of breast skin elasticity and breast skin thickness. Ski Res Technol 19:e191–e199. https://doi.org/10.1111/j.1600-0846.2012.00627.x
8. Arroyo J, Saavedra AC, Guerrero J, Montenegro P, Aguilar J, Pinto JA et al (2018) Breast elastography: identification of benign and malignant cancer based on absolute elastic modulus measurement using vibro-elastography. In: Image perception, observer performance, and technology assessment, vol 10577, Proceeding of SPIE 10577, p 48. https://doi.org/10.1117/12.2293664
9. Bazira PJ, Ellis H, Mahadevan V (2022) Anatomy and physiology of the breast. Surg 40:79–83. https://doi.org/10.1016/J.MPSUR.2021.11.015
10. Gefen A, Dilmoney B (2007) Mechanics of the normal woman's breast. Technol Heal Care 15:259–271. https://doi.org/10.3233/thc-2007-15404
11. Van Houten EEW, Doyley MM, Kennedy FE, Weaver JB, Paulsen KD (2003) Initial in vivo experience with steady-state subzone-based MR elastography of the human breast. J Magn Reson Imaging 17:72–85. https://doi.org/10.1002/jmri.10232

12. Egorov V, Sarvazyan AP (2008) Mechanical imaging of the breast. IEEE Trans Med Imaging 27:1275–1287. https://doi.org/10.1109/TMI.2008.922192
13. Hagl DM, Popovic D, Hagness SC, Booske JH, Okoniewski M (2003) Sensing volume of open-ended coaxial probes for dielectric characterization of breast tissue at microwave frequencies. IEEE Trans Microw Theory Tech 51:1194–1206. https://doi.org/10.1109/TMTT.2003.809626
14. Chanda A, Singh G (2023) Introduction to human tissues. Mater Horiz Nat Nanomater 1–12. https://doi.org/10.1007/978-981-99-2225-3_1/COVER
15. McKnight AL, Kugel JL, Rossman PJ, Manduca A, Hartmann LC, Ehman RL (2002) MR elastography of breast cancer: preliminary results. Am J Roentgenol 178:1411–1417. https://doi.org/10.2214/AJR.178.6.1781411/ASSET/IMAGES/06_AB1356_09_LA.JPEG
16. Eder M, Raith S, Jalali J, Volf A, Settles M, Machens HG et al (2014) Comparison of different material models to simulate 3-D breast deformations using finite element analysis. Ann Biomed Eng 42:843–857. https://doi.org/10.1007/S10439-013-0962-8/FIGURES/7
17. Bercoff J, Tanter M, Fink M (2004) Supersonic shear imaging: a new technique for soft tissue elasticity mapping. IEEE Trans Ultrason Ferroelectr Freq Control 51:396–409. https://doi.org/10.1109/TUFFC.2004.1295425
18. Cox TR, Erler JT (2011) Remodeling and homeostasis of the extracellular matrix: implications for fibrotic diseases and cancer. DMM Dis Model Mech 4:165–178. https://doi.org/10.1242/dmm.004077
19. Unlu MZ, Krol A, Magri A, Mandel JA, Lee W, Baum KG et al (2010) Computerized method for nonrigid MR-to-PET breast-image registration. Comput Biol Med 40:37–53. https://doi.org/10.1016/j.compbiomed.2009.10.010
20. Samani A, Zubovits J, Plewes D (2007) Elastic moduli of normal and pathological human breast tissues: an inversion-technique-based investigation of 169 samples. Phys Med Biol 52:1565–1576. https://doi.org/10.1088/0031-9155/52/6/002
21. Boyd NF, Martin LJ, Bronskill M, Yaffe MJ, Duric N, Minkin S (2010) Breast tissue composition and susceptibility to breast cancer. JNCI J Natl Cancer Inst 102:1224–1237. https://doi.org/10.1093/JNCI/DJQ239
22. Tanter M, Bercoff J, Athanasiou A, Deffieux T, Gennisson JL, Montaldo G et al (2008) Quantitative assessment of breast lesion viscoelasticity: initial clinical results using supersonic shear imaging. Ultrasound Med Biol 34:1373–1386. https://doi.org/10.1016/J.ULTRASMEDBIO.2008.02.002
23. Hoyt K, Castaneda B, Zhang M, Nigwekar P, di Sant'Agnese PA, Joseph JV et al (2008) Tissue elasticity properties as biomarkers for prostate cancer. Cancer Biomarkers 4:213–225. https://doi.org/10.3233/CBM-2008-44-505
24. Singh G, Chanda A (2021) Mechanical properties of whole-body soft human tissues: a review. Biomed Mater 16:062004. https://doi.org/10.1088/1748-605X/AC2B7A
25. Singh G, Gupta V, Chanda A (2022) Artificial skin with varying biomechanical properties. Mater Today Proc 62:3162–3166. https://doi.org/10.1016/J.MATPR.2022.03.433
26. Gupta V, Singla R, Singh G, Chanda A (2023) Development of soft composite based anisotropic synthetic skin for biomechanical testing. Fibers 11:55. https://doi.org/10.3390/FIB11060055
27. Makode S, Singh G, Chanda A (2021) Development of novel anisotropic skin simulants. Phys Scr 96:125019. https://doi.org/10.1088/1402-4896/AC2EFD
28. Chanda A, Singh G (2023) Applications, challenges, and future opportunities. Mater Horiz Nat Nanomater 285–292. https://doi.org/10.1007/978-981-99-2225-3_8/COVER
29. Singh G, Chanda A (2023) Development and mechanical characterization of artificial surrogates for brain tissues. Biomed Eng Adv 5:100084. https://doi.org/10.1016/J.BEA.2023.100084
30. Park JH, Park JU, Chang H (2021) Advances in biomaterials for breast reconstruction. Appl Sci 11:7493. https://doi.org/10.3390/APP11167493
31. Singh G, Chanda A (2021) Mechanical properties of whole-body soft human tissues: a review. Biomed Mater 16:062004

Chapter 9
Liver Tissue Simulants

When compared to the skin, the liver is the second biggest organ in the body. It is situated on the right side of the abdomen and weighs around three pounds. The liver is one of the visceral organs in the human body. It is responsible for a wide range of processes that are both essential and complex. These functions include the metabolism of proteins, carbohydrates, and lipids, as well as the storage of minerals and vitamins [1]. Most importantly, the liver is responsible for the generation of bile and the absorption of that bile into the small intestine, both of which contribute to the efficient digestion of food. There are a great number of endothelial cells, vascular cells, and parenchymal cells that make up the tissue of the liver [2, 3]. These cells include epithelial cells, stellate cells, hepatocytes, and more.

9.1 Structure of Liver Tissue

Hepatocytes are the primary functional cells of the liver, constituting about 70–85% of the liver's cellular mass. They play a central role in various metabolic processes and detoxification. Sinusoids are specialized blood vessels within the liver that allow the exchange of materials between blood and hepatocytes. They are lined with endothelial cells. Kupffer cells are specialized macrophages located within the sinusoids. They contribute to the immune function of the liver by phagocytosing foreign particles and microorganisms. Bile canaliculi, nearer to the hepatocytes, are crucial for the digestion of food [1, 4]. They are tiny ducts formed by adjacent hepatocytes. They collect bile produced by hepatocytes and transport it toward the bile ducts. Lined with epithelial cells, bile ducts are tubular structures that collect and transport bile from the liver to the gallbladder or directly to the small intestine. The liver is supported by connective tissue that forms its stroma. This tissue provides structural support to the liver and contains blood vessels and ducts [5, 6].

A. Chanda and G. Singh, *Soft Tissue Simulants*, Biomedical Materials for Multi-functional Applications, https://doi.org/10.1007/978-981-97-3060-5_9

9.1.1 Functions of Liver Tissue

The liver helps regulate blood glucose levels by storing glucose as glycogen and converting glycogen back to glucose as needed. It is involved in lipid synthesis, storage, and breakdown. The liver produces and secretes lipoproteins. The liver plays a crucial role in detoxifying harmful substances, including drugs and toxins, by transforming them into water-soluble compounds for excretion [7, 8]. Hepatocytes produce bile, which is essential for the digestion and absorption of fats. Bile is stored in the gallbladder and released into the small intestine as needed. The liver stores essential nutrients such as vitamins and minerals. It also stores glycogen, which can be broken down into glucose to meet energy needs. Kupffer cells within the liver are part of the immune system, helping to remove and destroy bacteria, viruses, and other foreign particles that may enter the bloodstream through the gastrointestinal tract. The liver regulates blood composition by removing excess nutrients and hormones. It also plays a role in blood clotting by producing clotting factors [9].

9.2 Traditional Liver Tissue for Testing

Before the development of advanced in vitro and ex vivo models, traditional liver tissues were commonly used for testing and research purposes. These traditional models included whole animal models, animal liver slices, and primary hepatocyte cultures. While these models have contributed valuable insights, they come with limitations. The use of intact animals, such as rodents (mice, rats), for studying liver function and responses to drugs or toxins is most common across all the traditional methods [10]. They are widely used for assessing overall physiological effects on the liver and studying the metabolism, toxicity, and pharmacokinetics of drugs. Animal liver slices are the second traditionally used tissue samples for drug delivery and other stimuli approach [1, 11, 12]. Thin slices of liver tissue were obtained from animals, typically rodents, for ex vivo experiments. These thin slices are used for assessing drug metabolism and toxicity and studying tissue responses to various stimuli. The liver is perfused with a nutrient solution, allowing ex vivo studies of liver function. Liver perfusion models allowed studying metabolism and drug clearance.

To date, in order to assess the mechanical properties of liver tissue, porcine and bovine animal models have been used [7, 11, 13, 14] (see Fig. 9.1). As a result of the fact that the liver tissue is one of the organs that is most often damaged by trauma injuries, the mechanical characteristics of the liver tissue are of significant interest. In light of this fact, it is envisaged that the experimental data of the cadaveric liver tissue will be used for the study of trauma studies and damage biomechanics under settings that are representative of actual mechanical loading requirements [15].

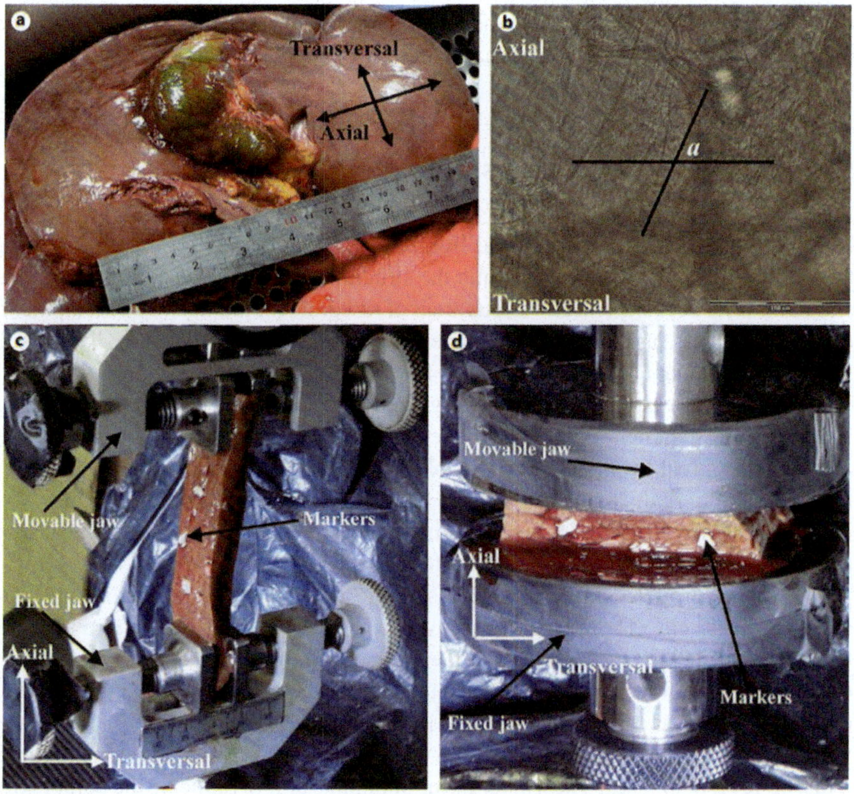

Fig. 9.1 **a** Fresh human cadaveric liver tissues; **b** 3 μm thick histological section with the axial and transversal orientations was prepared; **c** tensile test as well as the **d** compressive test were carried out and the markers were tracked using the CCD cameras [15]

9.2.1 Challenges with the Traditional Liver Tissue

The primary limitation of traditional liver tissue models is species differences, i.e., findings from animal models may not always directly translate to human biology due to species differences in metabolism and physiology. Second, obtaining fresh and healthy liver tissue from animals may be challenging, and the supply is limited. Third, variability in the genetic makeup of animals or donors can introduce inconsistency in experimental results. Fourth, animals and animal tissues may not accurately reflect human-specific responses to drugs or diseases [16–18]. Fifth, the use of animals for experimentation raises ethical concerns related to animal welfare. Recent advancements have led to the development of more sophisticated liver tissue models, including human cell-derived liver organoids, 3D liver cell cultures, and microphysiological systems (liver-on-a-chip). These advanced models aim to better mimic the complexity of the human liver, allowing for more accurate predictions of drug metabolism, toxicity, and disease responses. They also address some of the limitations associated with traditional liver tissue models [15, 19].

9.3 Liver Tissue Simulants

Liver tissue simulants, also known as liver tissue models or in vitro liver models, are artificial constructs designed to mimic the structure and function of the liver. These simulants play a crucial role in various areas of research, including drug development, disease modeling, and understanding liver physiology. Creating effective liver tissue simulants is crucial for reducing the need for animal testing and advancing our understanding of liver-related diseases and drug interactions [20, 21]. Using 3D bioprinting technology, researchers can create three-dimensional structures that resemble the architecture of real liver tissue. This involves layering cells and supportive materials to create a more realistic tissue structure [21, 22] (see Fig. 9.2). Materials like polymers can be engineered to mimic certain aspects of liver tissue, such as elasticity and porosity. These synthetic models offer reproducibility and control over specific properties. Hydrogels can be used as a scaffold to support the growth and organization of liver cells. These scaffolds provide a three-dimensional structure for cells to grow and interact more naturally. Computational models play a crucial role in the design and development of liver tissue simulants by providing insights into the complex biological processes involved [10, 23]. These models can help researchers understand the behavior of liver tissue at various scales, from molecular interactions to tissue-level responses. These liver tissue simulants aim to replicate the complexity of the liver, allowing researchers to study physiological responses, metabolism, and toxicity in a controlled environment. The choice of simulant depends on the specific research goals and the desired level of complexity and realism needed for the study [24].

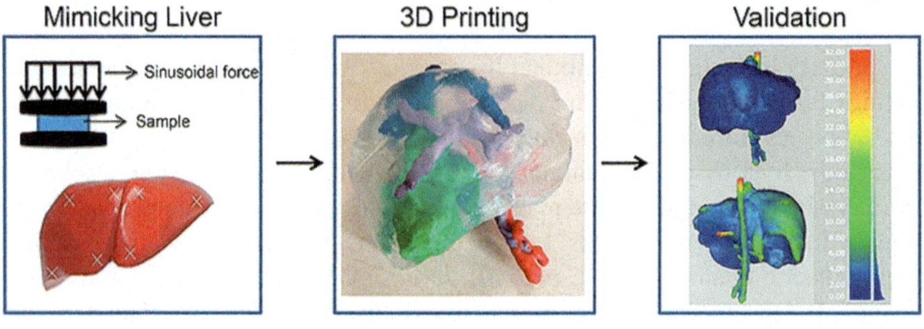

Fig. 9.2 Mimicking the liver tissue using 3D printing technology [25]

9.4 Types of Liver Tissue Simulants

Liver tissue simulants aim to mimic the structure and function of real liver tissue for various applications in research, drug testing, and medical device development. The choice of a specific liver tissue simulant depends on the intended use and the level of complexity required for the study. To study drug delivery and drug metabolism, 2D and 3D cell culture models are used. Hepatocytes and other relevant liver cell types are cultured on flat surfaces (e.g., Petri dishes) to study basic cellular functions, drug metabolism, and toxicity [10, 21, 26]. Their primary function is the initial screening of drug candidates and studying basic cellular processes. Cells are cultured in three-dimensional environments, allowing for better mimicry of in vivo conditions. Spheroids, organoids, or hydrogel-based cultures are common. They allow improved representation of tissue architecture, drug metabolism, and cell–cell interactions. 3D bioprinting technology is used to create layered structures of liver cells and supportive materials, mimicking the complexity of liver tissue [10, 26]. These simulants and structures are useful for tissue engineering, drug testing, and personalized medicine applications.

Polymers are engineered to replicate certain physical and chemical properties of liver tissue. These models can be synthetic or natural polymers and used for customization of tissue properties, drug testing, and material compatibility studies. Artificial materials are used to mimic specific aspects of liver tissue, such as elasticity and porosity. They allow controlled studies of tissue properties, drug testing, and material compatibility assessments [27–29]. Computational models simulate the behavior of liver tissue at various levels, including cellular processes, metabolism, and tissue mechanics. These models are appropriate for predicting drug metabolism, understanding tissue responses, and optimizing experimental designs. To study liver physiology, drug metabolism, and testing therapeutic interventions, portions of liver tissue are maintained in perfusion systems that mimic physiological conditions, including blood flow [30–32].

The choice of liver tissue simulant depends on the specific research goals, the level of complexity required, and the intended applications. Researchers often use a combination of these models to gain a more comprehensive understanding of liver tissue behavior in different contexts.

9.5 Why Liver Tissue Simulants Are in Trend?

Liver tissue simulants are gaining popularity and are considered "in trend" for several reasons, reflecting advancements in technology, ethical considerations, and the need for more efficient and reliable methods in various scientific and medical fields. Liver tissue simulants offer alternatives to traditional animal testing, aligning with ethical concerns about the use of animals in research [10, 26]. This shift is in line with a growing awareness of animal welfare and a desire to reduce the number of animals used in experiments. Technological advancements, such as 3D bioprinting and tissue engineering techniques, have enabled the creation of more sophisticated and biomimetic liver tissue simulants. These technologies allow researchers to replicate the complex structure and function of the liver more accurately [21]. Liver tissue simulants can be customized to mimic the characteristics of individual patients, supporting the growing field of precision medicine. This allows for personalized drug testing and treatment strategies based on the specific genetic and physiological makeup of an individual. Liver tissue simulants provide a faster and more cost-effective way to test the safety and efficacy of drugs in the early stages of development. This can accelerate the drug development process and reduce the overall cost of bringing new drugs to market [1, 10]. Liver tissue simulants allow for the early detection of potential drug toxicities, reducing the likelihood of adverse effects during clinical trials. This can improve the safety profile of new drugs before they enter human testing. Regulatory agencies are increasingly accepting and validating the use of liver tissue simulants in certain preclinical testing scenarios. This acceptance has encouraged researchers and pharmaceutical companies to adopt these models as part of their drug development strategies.

In summary, the current trend toward liver tissue simulants is driven by a combination of ethical considerations, technological advancements, efficiency in drug development, and the desire to achieve more personalized and precise medical interventions. As these models continue to evolve, their applications are likely to expand further, contributing to advancements in medicine and biology.

9.6 Applications of Liver Tissue Simulants

Liver tissue simulants find applications across various fields, contributing to advancements in medical research, drug development, and understanding of liver-related diseases. Liver tissue simulants are used to study drug metabolism, including the breakdown of drugs by liver enzymes. This helps predict how drugs will be processed in the human body. Simulants enable the evaluation of drug toxicity, helping to identify potential adverse effects on the liver and improving safety profiles before clinical trials. Simulants allow researchers to model and study various liver diseases, including viral hepatitis, fatty liver disease, and cirrhosis [24]. This aids in understanding disease mechanisms and developing potential treatments. Liver tissue simulants contribute to the development of functional liver tissue for transplantation, reducing the reliance on donor organs and addressing organ shortage issues. Simulants help assess the compatibility of biomaterials with liver tissue, essential for designing implants and devices [33–36]. Simulants can be tailored to mimic the unique characteristics of an individual's liver tissue, allowing for personalized drug testing and treatment strategies. Simulants enable the study of patient-specific disease mechanisms, facilitating the development of targeted therapies based on individual variations. Simulants allow for controlled experiments to understand how the liver responds to various stimuli, such as drugs, toxins, and environmental factors [29, 37,

38]. Integration of liver tissue simulants with other organ models on a chip enables the study of inter-organ interactions and systemic effects of drugs. Simulants facilitate detailed studies of cellular and molecular processes within the liver, aiding in the discovery of new biological pathways and drug targets.

Computational models using liver tissue simulants assist in predicting drug behavior in the body, guiding the optimization of drug dosages and administration schedules. Liver tissue simulants serve as educational tools for medical professionals, providing realistic training environments for procedures such as liver biopsies and surgeries. Researchers and students can use simulants to understand the drug development process and the impact of drugs on liver function [29, 32, 37–41]. Overall, liver tissue simulants play a pivotal role in advancing our understanding of liver biology, improving drug development processes, and developing innovative medical solutions with broad implications for human health.

References

1. Yeh WC, Li PC, Jeng YM, Hsu HC, Kuo PL, Li ML et al (2002) Elastic modulus measurements of human liver and correlation with pathology. Ultrasound Med Biol 28:467–474. https://doi.org/10.1016/S0301-5629(02)00489-1
2. Kazemnejad S (2009) Hepatic tissue engineering using scaffolds: state of the art. Avicenna J Med Biotechnol 1:135–145
3. Lorenzini S, Andreone P (2007) Stem cell therapy for human liver cirrhosis: a cautious analysis of the results. Stem Cells 25:2383–2384. https://doi.org/10.1634/stemcells.2007-0056
4. Wells RG (2013) Tissue mechanics and fibrosis. Biochim Biophys Acta—Mol Basis Dis 1832:884–890. https://doi.org/10.1016/J.BBADIS.2013.02.007
5. Evans DW, Moran EC, Baptista PM, Soker S, Sparks JL (2013) Scale-dependent mechanical properties of native and decellularized liver tissue. Biomech Model Mechanobiol 12:569–580. https://doi.org/10.1007/s10237-012-0426-3
6. Nava A, Mazza E, Kleinermann F, Avis NJ, McClure J, Bajka M (2004) Evaluation of the mechanical properties of human liver and kidney through aspiration experiments. Technol Heal Care 12:269–280. https://doi.org/10.3233/thc-2004-12306
7. Hollenstein M, Nava A, Valtorta D, Snedeker JG, Mazza E (2006) Mechanical characterization of the liver capsule and parenchyma. In: Lecture notes computer science (including Subseries Lecture notes in artificial intelligence (LNAI) and Lecture notes in bioinformatics), vol 4072, LNCS. Springer, pp 150–158. https://doi.org/10.1007/11790273_17
8. Umale S, Chatelin S, Bourdet N, Deck C, Diana M, Dhumane P et al (2011) Experimental in vitro mechanical characterization of porcine Glisson's capsule and hepatic veins. J Biomech 44:1678–1683. https://doi.org/10.1016/j.jbiomech.2011.03.029
9. Stingl J, Báĉa V, Ĉech P, Kovanda J, Kovandová H, Mandys V et al (2002) Morphology and some biomechanical properties of human liver and spleen. Surg Radiol Anat 24:285–289. https://doi.org/10.1007/s00276-002-0054-1
10. Cui H, Nowicki M, Fisher JP, Zhang LG (2017) 3D Bioprinting for organ regeneration. Adv Healthc Mater 6:1601118. https://doi.org/10.1002/ADHM.201601118
11. Brunon A, Bruyère-Garnier K, Coret M (2010) Mechanical characterization of liver capsule through uniaxial quasi-static tensile tests until failure. J Biomech 43:2221–2227. https://doi.org/10.1016/j.jbiomech.2010.03.038
12. Mattei G, Ahluwalia A (2016) Sample, testing and analysis variables affecting liver mechanical properties: a review. Acta Biomater 45:60–71. https://doi.org/10.1016/j.actbio.2016.08.055

13. Chui C, Kobayashi E, Chen X, Hisada T, Sakuma I (2007) Transversely isotropic properties of porcine liver tissue: experiments and constitutive modelling. Med Biol Eng Comput 45:99–106. https://doi.org/10.1007/s11517-006-0137-y
14. Lu YC, Kemper AR, Untaroiu CD (2014) Effect of storage on tensile material properties of bovine liver. J Mech Behav Biomed Mater 29:339–349. https://doi.org/10.1016/j.jmbbm.2013.09.022
15. Karimi A, Shojaei A (2018) An experimental study to measure the mechanical properties of the human liver. Dig Dis 36:150–155. https://doi.org/10.1159/000481344
16. Gungor-Ozkerim PS, Inci I, Zhang YS, Khademhosseini A, Dokmeci MR (2018) Bioinks for 3D bioprinting: an overview. Biomater Sci 6:915–946. https://doi.org/10.1039/C7BM00765E
17. Herrera J, Henke CA, Bitterman PB (2018) Extracellular matrix as a driver of progressive fibrosis. J Clin Invest 128:45–53. https://doi.org/10.1172/JCI93557
18. Manduca A, Oliphant TE, Dresner MA, Mahowald JL, Kruse SA, Amromin E et al (2001) Magnetic resonance elastography: non-invasive mapping of tissue elasticity. Med Image Anal 5:237–254. https://doi.org/10.1016/S1361-8415(00)00039-6
19. Huwart L, Peeters F, Sinkus R, Annet L, Salameh N, ter Beek LC et al (2006) Liver fibrosis: non-invasive assessment with MR elastography. NMR Biomed 19:173–179. https://doi.org/10.1002/nbm.1030
20. Venkatesh SK, Yin M, Ehman RL (2013) Magnetic resonance elastography of liver: technique, analysis, and clinical applications. J Magn Reson Imaging 37:544–555. https://doi.org/10.1002/JMRI.23731
21. Jammalamadaka U, Tappa K (2018) Recent advances in biomaterials for 3D printing and tissue engineering. J Funct Biomater 9:22. https://doi.org/10.3390/JFB9010022
22. Matai I, Kaur G, Seyedsalehi A, McClinton A, Laurencin CT (2020) Progress in 3D bioprinting technology for tissue/organ regenerative engineering. Biomaterials 226:119536. https://doi.org/10.1016/J.BIOMATERIALS.2019.119536
23. Huang W-H, Chui C-K, Kobayashi E, Teoh S-H, Chang S, Huang W-H et al (2010) Multi-scale model for investigating the electrical properties and mechanical properties of liver tissue undergoing ablation. Int J Comput Assist Radiol Surg 6(5):601–607. https://doi.org/10.1007/S11548-010-0537-1
24. Singh G, Chanda A (2021) Mechanical properties of whole-body soft human tissues: a review. Biomed Mater 16:062004. https://doi.org/10.1088/1748-605X/AC2B7A
25. Tejo-Otero A, Lustig-Gainza P, Fenollosa-Artés F, Valls A, Krauel L, Buj-Corral I (2020) 3D printed soft surgical planning prototype for a biliary tract rhabdomyosarcoma. J Mech Behav Biomed Mater 109:103844. https://doi.org/10.1016/J.JMBBM.2020.103844
26. Vijayavenkataraman S, Yan WC, Lu WF, Wang CH, Fuh JYH (2018) 3D bioprinting of tissues and organs for regenerative medicine. Adv Drug Deliv Rev 132:296–332. https://doi.org/10.1016/J.ADDR.2018.07.004
27. Makode S, Singh G, Chanda A (2021) Development of novel anisotropic skin simulants. Phys Scr 96:125019. https://doi.org/10.1088/1402-4896/AC2EFD
28. Chanda A, Unnikrishnan V, Lackey K, Robbins J (2020) Biofidelic conductive soft tissue surrogates. Int J Polym Mater Polym Biomater 69:127–135. https://doi.org/10.1080/00914037.2018.1552856
29. Gupta V, Singh G, Chanda A (2023) Development of novel hierarchical designs for skin graft simulants with high expansion potential. Biomed Phys Eng Express 9:035024. https://doi.org/10.1088/2057-1976/ACC661
30. Gupta V, Singla R, Singh G, Chanda A (2023) Development of soft composite based anisotropic synthetic skin for biomechanical testing. Fibers 11:55. https://doi.org/10.3390/FIB11060055
31. Chanda A, Singh G (2023) Tissues in functional organs—low stiffness. Mater Horiz Nat Nanomater 33–48. https://doi.org/10.1007/978-981-99-2225-3_4/COVER
32. Chanda A, Singh G (2023) Applications, challenges, and future opportunities. Mater Horiz Nat Nanomater 85–92. https://doi.org/10.1007/978-981-99-2225-3_8/COVER
33. Gupta V, Singh G, Chanda A (2022) Development and testing of skin grafts models with varying slit orientations. Mater Today Proc 62:3462–3467. https://doi.org/10.1016/J.MATPR.2022.04.282

34. Gupta V, Chanda A (2022) Biomechanics of skin grafts: effect of pattern size, spacing and orientation. Eng Res Express 4:015006. https://doi.org/10.1088/2631-8695/AC48CB
35. Gupta V, Chanda A (2022) Expansion potential of skin grafts with alternating slit based auxetic incisions. Forces Mech 7:100092. https://doi.org/10.1016/J.FINMEC.2022.100092
36. Chhikara K, Singh G, Gupta S, Chanda A (2022) Progress of additive manufacturing in fabrication of foot orthoses for diabetic patients: a review. Ann 3D Print Med 8:100085. https://doi.org/10.1016/J.STLM.2022.100085
37. Gupta V, Singh G, Chanda A (2023) High expansion auxetic skin graft simulants for severe burn injury mitigation. Eur Burn J 4:108–20. https://doi.org/10.3390/EBJ4010011
38. Gupta V, Singh G, Gupta S, Chanda A (2023) Expansion potential of auxetic prosthetic skin grafts: a review. Eng Res Express 5:022003. https://doi.org/10.1088/2631-8695/ACCFE5
39. Singh G, Chanda A (2023) Development and biomechanical testing of human stomach tissue surrogates. Mater Horiz Nat Nanomater Part F1471:113–125. https://doi.org/10.1007/978-981-99-5064-5_7/COVER
40. Singh G, Chanda A (2023) Biofidelic tongue and tonsils tissue surrogates. Mater Horiz Nat Nanomater Part F1471:159–170. https://doi.org/10.1007/978-981-99-5064-5_10/COVER
41. Chanda A, Singh G (2023) Introduction to human tissues. Mater Horiz Nat Nanomater 1–12. https://doi.org/10.1007/978-981-99-2225-3_1/COVER

Chapter 10
Stomach Tissue Simulants

The stomach is a muscular organ located in the upper abdomen, and its primary function is to break down and digest food. The stomach tissue is composed of several layers, each with distinct structures and functions [1]. The innermost layer of the stomach is called the mucosa, and it is lined by a specialized epithelium known as the gastric epithelium. This epithelium contains various cell types, including mucous cells, parietal cells, chief cells, and endocrine cells. Gastric glands are invaginations in the mucosa that extend into the underlying layers. They secrete gastric juices containing enzymes and acids that aid in the digestion of food [2, 3]. The submucosa is a layer of connective tissue beneath the mucosa. It contains blood vessels, lymphatic vessels, and nerves that support the function of the stomach. The muscularis externa is responsible for the peristaltic contractions that propel food through the stomach during digestion. It consists of three layers of smooth muscle: an inner oblique layer, a middle circular layer, and an outer longitudinal layer. The outermost layer of the stomach is the serosa. It is composed of connective tissue and a layer of simple squamous epithelium. The serosa provides protection and support for the stomach [2, 4, 5]. A schematic image of the stomach tissue is shown in Fig. 10.1. This figure highlights the three primary sections of the stomach, which are the antrum, corpus, and fundus, as well as the duodenum, lower esophageal sphincter, and other structures [6].

The stomach undergoes dynamic changes during the digestive process. When food enters the stomach, the muscular walls contract to mix and grind the food, creating a semi-liquid mixture known as chyme. The gastric glands in the mucosa release digestive enzymes and acids to help break down proteins and other components of the food. Additionally, the stomach is equipped with protective mechanisms to prevent damage from its own digestive juices [7]. The mucous cells in the gastric epithelium secrete mucus, forming a protective layer that lines the stomach and shields it from the corrosive effects of stomach acid. The study of stomach tissue is crucial for understanding digestive processes and gastric diseases and developing treatments for conditions such as peptic ulcers, gastritis, and gastric cancer. Researchers may

A. Chanda and G. Singh, *Soft Tissue Simulants*, Biomedical Materials for Multi-functional Applications, https://doi.org/10.1007/978-981-97-3060-5_10

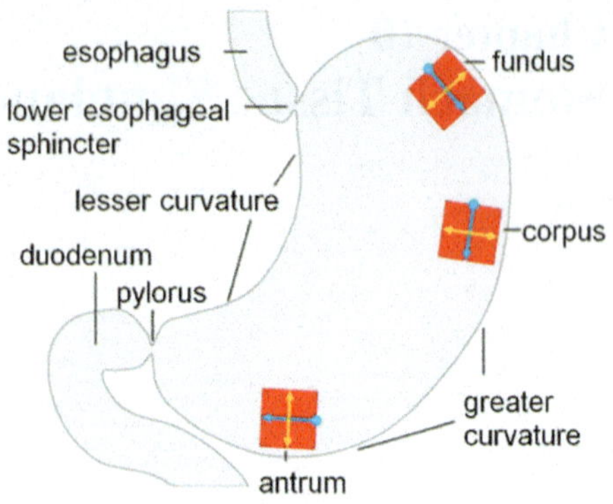

Fig. 10.1 Schematic illustration of the stomach tissue [6]

use various techniques, including histology, immunohistochemistry, and molecular biology, to investigate the structure and function of stomach tissue [2, 8, 9].

10.1 Scientific Research Areas of Stomach Tissue

Stomach tissue, both in its natural state and as part of tissue-engineered models, has various applications in scientific research, medical diagnostics, and therapeutic development. Stomach tissue is used to study the mechanical aspects of digestion, including the peristaltic contractions that mix and propel food through the digestive tract. Researchers investigate the secretion of gastric juices, including acids and enzymes, to understand how stomach tissue contributes to the digestive process. Stomach tissue is utilized for in vitro testing of drug toxicity, helping to assess the potential adverse effects of pharmaceutical compounds on stomach cells and tissues [7, 10]. Stomach tissue models aid in studying the pathophysiology of conditions like gastritis and peptic ulcers, including the impact of bacteria (such as Helicobacter pylori) and the role of inflammation. Investigating stomach tissue contributes to understanding the development and progression of gastric cancer, facilitating the development of targeted therapies. Researchers use stomach cells to create bioengineered tissues for transplantation or replacement of damaged stomach tissue. Stomach tissue serves as a source for creating scaffolds that support the growth of cells in tissue engineering applications [9, 11, 12].

Stomach tissue studies aid in designing drug delivery systems that can release medications at specific rates or locations within the stomach. Stomach tissue models are used for training medical professionals, particularly in procedures related to gastrointestinal surgery, such as gastrectomy. Simulated stomach tissues can be employed for training in endoscopic procedures, allowing medical practitioners to enhance their skills in a controlled environment. Investigating stomach tissue helps in understanding the mechanisms of nutrient absorption from the stomach, contributing to our knowledge of overall nutrient uptake in the digestive system [11, 13]. Stomach tissue may be integrated into biosensing platforms for the development of diagnostic tools that can detect specific biomarkers associated with gastrointestinal diseases. Studying stomach tissue contributes to the identification of potential biomarkers for diagnosing stomach-related conditions. Stomach tissue models can be personalized using patient-derived cells, enabling researchers to study disease mechanisms and test personalized therapeutic approaches. These applications demonstrate the diverse ways in which stomach tissue contributes to advancing scientific knowledge, medical treatments, and healthcare practices. Researchers continue to explore innovative approaches to leverage stomach tissue for improved diagnostics, therapeutics, and tissue engineering strategies [14–17].

10.2 Traditional Stomach Tissues

Traditional stomach tissues typically refer to tissue samples obtained from the stomach, either from humans or animals, for various research and medical purposes. These tissues are commonly used in histological, physiological, and pathological studies to understand the structure, function, and diseases of the stomach. Tissue samples obtained during endoscopic procedures are common in clinical practice. These biopsies allow clinicians to diagnose and monitor various stomach conditions, including inflammation, ulcers, and tumors [2, 3, 14]. Stomach tissues obtained during post-mortem examinations provide insights into the gross and histological characteristics of the stomach. Autopsy findings can be valuable for understanding the pathological aspects of stomach diseases. Tissues removed during stomach surgeries, such as gastrectomies, may be collected for pathological examination [12, 18]. These specimens help assess the extent of disease, margins, and the effectiveness of surgical interventions. Tissues from animals, such as rodents or pigs, are commonly used in research to study basic physiological processes, test experimental drugs, and model human stomach conditions. These tissues can be obtained post-mortem or through specific procedures [2, 19].

Cultured cells derived from stomach tissues are used in in vitro studies. These cell lines, maintained in laboratory conditions, provide a controlled environment for experiments related to drug testing, cellular responses, and molecular biology. In cases of organ transplantation, portions of the stomach from donors may be used to replace damaged or diseased stomach tissue in recipients. Stomach tissues collected from patients and healthy individuals may be preserved in biobanks for future research [8, 12]. These repositories serve as valuable resources for studying various aspects of stomach biology and diseases. Traditional stomach tissues play a crucial role in advancing our understanding of stomach biology, diseases, and therapeutic interventions. However, as research techniques continue to evolve, there is a growing emphasis on combining traditional tissue approaches with cutting-edge technologies, such as organoids and organ-on-a-chip systems, to better model the complexity of the stomach microenvironment in vitro [17, 20].

10.2.1 Challenges with Traditional Stomach Tissues

Traditional approaches to studying and treating stomach tissues face several challenges, particularly in the context of gastrointestinal disorders, surgeries, and general research. Creating accurate in vitro models of the stomach is challenging. Traditional cell cultures often lack the three-dimensional complexity and dynamic physiological conditions present in the native stomach environment [5, 12, 21]. The stomach is a complex organ with various cell types, including mucous-secreting cells, parietal cells, chief cells, and enteroendocrine cells. Replicating this cellular diversity in in vitro models remains a significant challenge. Gastrointestinal diseases, such as peptic ulcers, gastritis, and gastric cancer, pose challenges for researchers. Developing accurate disease models that mimic the complexity of these conditions is crucial for understanding disease mechanisms and testing potential therapies [17, 22]. Studying drug absorption in the stomach is challenging due to the harsh acidic environment. Traditional methods may not accurately reflect the in vivo conditions, impacting the prediction of drug behavior and efficacy. Surgical interventions involving the stomach, such as gastric bypass surgeries, can lead to complications. These may include postoperative leaks, infections, and issues related to altered anatomy. Overcoming these complications requires a comprehensive understanding of stomach tissue response to surgery [2, 3].

Studying and replicating the complex motility patterns of the stomach is difficult. Traditional models often fail to capture the dynamic peristaltic movements that are essential for digestion and nutrient absorption. Identifying reliable biomarkers for stomach-related diseases is a challenge. Robust biomarkers are crucial for early diagnosis, prognosis, and monitoring of treatment responses. Regenerating damaged or diseased stomach tissue remains a significant challenge. Developing strategies for tissue engineering and regeneration is crucial for treating conditions such as ulcers or injuries. Conducting research on stomach tissues may involve ethical concerns, especially when studying human tissues [6, 23–25]. Obtaining tissue samples and conducting experiments raise ethical considerations. Additionally, technical challenges in maintaining the viability of tissues and ensuring reproducibility are common.

Addressing these challenges requires a multidisciplinary approach that combines advances in tissue engineering, organoids, microfluidics, and other cutting-edge technologies. Improved in vitro and in vivo models, along with a deeper understanding of stomach physiology and pathology, will contribute to advancements in the diagnosis and treatment of stomach-related disorders [2].

10.3 Stomach Tissues Simulants

Stomach tissue simulants are artificial models designed to mimic the structural and functional properties of natural stomach tissues. These simulants serve as valuable tools in scientific research, drug development, and medical education. Creating stomach tissue simulants involves replicating key features of the stomach, such as its cellular composition, mechanical properties, and physiological functions. Several approaches have been explored to create stomach tissue simulants, each with its advantages and challenges [11, 19, 20].

10.3.1 Types of Stomach Tissue Simulants

Various polymers, such as silicone elastomers or hydrogels, are used to create stomach tissue simulants. These materials can be engineered to mimic the mechanical properties and elasticity of natural stomach tissue. Silicone-based materials are widely employed for drug testing, studying mechanical responses of the stomach, and developing flexible models for endoscopic training. Biodegradable materials, such as certain polymers or extracellular matrix components, can be used to create scaffolds that support cell growth. These scaffolds can be populated with cells to create more complex stomach tissue simulants and can be used in tissue engineering, regenerative medicine, and drug testing with three-dimensional (3D) structures [11, 20, 26, 27]. Microfluidic devices with compartments represent different layers of the stomach, including the mucosa and muscularis externa. These devices can incorporate living cells to replicate the microenvironment of the stomach to study physiological responses, drug metabolism, and interactions between stomach tissue and pathogens.

Hydrogels, which can mimic the water content and flexibility of natural tissues, are used to create stomach tissue simulants. These models may incorporate cells and bioactive molecules and help in drug testing, studying cellular responses, and providing a platform for controlled drug release. Combining cells with biomaterials to create tissue constructs that replicate the cellular composition and mechanical properties of natural stomach tissues [8, 26, 28, 29].

10.3.2 Key Considerations in Stomach Tissue Simulant Development

There are some key considerations for the development of stomach tissue simulant. First, simulants should incorporate relevant cell types found in the stomach, including epithelial cells, mucous cells, and smooth muscle cells. Second, the mechanical properties should be the key consideration for the fabrication of stomach tissue simulant.

Mimicking the elasticity and compliance of natural stomach tissue to accurately replicate mechanical responses during digestion. Third, striving for biomimicry to create models that closely resemble the structure and function of natural stomach tissues. Fourth, ensuring the reproducibility of stomach tissue simulants to allow for consistent and reliable experimental results [2, 16, 30–36].

10.4 Applications of Stomach Tissue Simulants

Stomach tissue simulants, or in vitro models of the stomach, have various applications across research, drug development, and medical fields. These models provide a platform for studying stomach physiology, investigating disease mechanisms, and testing the effects of drugs. Stomach tissue simulants are used to study how drugs are absorbed, metabolized, and interact with stomach tissue. This aids in predicting drug behavior in vivo and optimizing dosages. Researchers can assess the effectiveness of drugs in treating stomach-related conditions and study potential adverse effects, helping to identify safer and more efficient therapeutic options. Stomach tissue simulants enable the study of various gastrointestinal diseases, such as gastritis, peptic ulcers, and gastric cancer [13, 17]. Researchers can mimic disease conditions to understand the underlying mechanisms and develop targeted treatments. Models can be infected with pathogens like Helicobacter pylori to study the infection process, immune responses, and potential interventions for bacterial-induced gastric diseases. Stomach tissue simulants contribute to the development of tissue-engineered constructs for potential transplantation or in situ regeneration. This involves using biomaterials and cells to recreate functional stomach tissue, offering new therapeutic options for conditions like ulcers or injuries [2].

Stomach tissue simulants provide educational tools for medical and research training. They offer a controlled environment for students and professionals to learn about stomach physiology, pathology, and experimental techniques. Stomach tissue simulants help in understanding the processes involved in nutrient absorption in the stomach. This is crucial for studying the effects of different diets, identifying factors influencing nutrient absorption, and developing interventions for malabsorption disorders. By utilizing patient-derived cells or tissues, stomach tissue simulants can be customized to reflect individual variations in drug responses, disease susceptibility, and genetic factors [37–39]. This approach holds promise for tailoring treatments to specific patient profiles. Stomach tissue simulants play a pivotal role in advancing our understanding of stomach biology, disease processes, and drug responses. As technologies continue to evolve, these models will likely become more sophisticated, contributing to further breakthroughs in gastroenterology and personalized medicine.

10.4.1 Challenges with Stomach Tissue Simulants

The stomach has a complex three-dimensional architecture with various cell types, including mucous-secreting cells, parietal cells, chief cells, and enteroendocrine cells. Mimicking this complexity in tissue simulants remains challenging. Replicating the dynamic physiological processes of the stomach, such as peristaltic movements, rhythmic contractions, and changes in pH, is difficult in static in vitro models. Recreating the mucosal barrier and its protective functions in stomach tissue simulants is crucial [11, 18]. Achieving a realistic mucosal layer, which protects against acid and digestive enzymes, is a technical challenge. Mimicking the physiological fluid dynamics of the stomach, including the controlled release of gastric juices, digestion, and nutrient absorption, is challenging in static in vitro systems [13]. Maintaining the viability and functionality of stomach tissue simulants over extended periods is crucial for chronic disease modeling and long-term drug testing. Establishing standardized protocols for the fabrication and maintenance of stomach tissue simulants is essential for ensuring reproducibility across different research studies and laboratories [14].

10.4.2 Future Directions for Stomach Tissue Simulants

First, development of more advanced three-dimensional models, such as stomach organoids or tissue-engineered constructs, that better mimic the native tissue architecture and cellular diversity. Second, continued development of microfluidic systems and organ-on-a-chip technologies to better replicate the dynamic physiological conditions of the stomach, including fluid flow and shear stress. Third, integration of advanced biomaterials that can enhance the structural integrity, mechanical properties, and functionality of stomach tissue simulants. Fourth, utilization of patient-derived cells or tissues to create more personalized and humanized stomach tissue simulants, accounting for individual variations in drug responses and disease susceptibility [2, 40, 41]. Fifth, integration of electrical stimulation to better replicate the electrical signaling and rhythmic contractions of the stomach. Sixth, development of interconnected multi-organ systems that simulate the crosstalk between the stomach and other organs, enabling a more holistic understanding of systemic effects. Collaborative efforts to establish standardized protocols for the fabrication, maintenance, and evaluation of stomach tissue simulants to improve reproducibility and comparability between studies. Continued development of ethical and regulatory frameworks that address the challenges associated with obtaining and using human stomach tissues for research while ensuring compliance with ethical standards and privacy concerns [40].

Advancements in stomach tissue simulants will likely involve a combination of technological innovations, interdisciplinary collaboration, and a deeper understanding of stomach physiology and pathology. These improvements will contribute to more accurate disease modeling, drug development, and personalized medicine in the field of gastroenterology.

References

1. Brandstaeter S, Fuchs SL, Aydin RC, Cyron CJ (2019) Mechanics of the stomach: a review of an emerging field of biomechanics. GAMM-Mitteilungen 42. https://doi.org/10.1002/gamm.201900001.
2. Singh G, Chanda A (2021) Mechanical properties of whole-body soft human tissues: a review. Biomed Mater 16:062004. https://doi.org/10.1088/1748-605X/AC2B7A
3. Chanda A, Singh G (2023) Introduction to human tissues. In: Materials horizons: from nature to nanomaterials, pp 1–12. https://doi.org/10.1007/978-981-99-2225-3_1/COVER
4. Jia ZG, Li W, Zhou ZR (2015) Mechanical characterization of stomach tissue under uniaxial tensile action. J Biomech 48:651–658. https://doi.org/10.1016/J.JBIOMECH.2014.12.048
5. Saraf H, Ramesh KT, Lennon AM, Merkle AC, Roberts JC (2007) Mechanical properties of soft human tissues under dynamic loading. J Biomech 40:1960–1967. https://doi.org/10.1016/J.JBIOMECH.2006.09.021
6. Aydin RC, Brandstaeter S, Braeu FA, Steigenberger M, Marcus RP, Nikolaou K et al (2017) Experimental characterization of the biaxial mechanical properties of porcine gastric tissue. J Mech Behav Biomed Mater 74:499–506. https://doi.org/10.1016/j.jmbbm.2017.07.028
7. Ní Annaidh A, Bruyère K, Destrade M, Gilchrist MD, Otténio M (2012) Characterization of the anisotropic mechanical properties of excised human skin. J Mech Behav Biomed Mater 5:139–148. https://doi.org/10.1016/J.JMBBM.2011.08.016
8. Huang J, Zhang Z, Jiang H (2023) Edible hydrogels with shrinkage tolerance in acids and stomach-friendly mechanical moduli. Appl Mater Today 32:101786. https://doi.org/10.1016/J.APMT.2023.101786
9. Luan Z, Liu S, Wang W, Xu K, Ye S, Dan R et al (2022) Aligned nanofibrous collagen membranes from fish swim bladder as a tough and acid-resistant suture for pH-regulated stomach perforation and tendon rupture. Biomater Res 26:1–20. https://doi.org/10.1186/S40824-022-00306-1/FIGURES/9
10. Takaza M, Moerman KM, Gindre J, Lyons G, Simms CK (2013) The anisotropic mechanical behaviour of passive skeletal muscle tissue subjected to large tensile strain. J Mech Behav Biomed Mater 17:209–220. https://doi.org/10.1016/J.JMBBM.2012.09.001
11. Shi Y, Glaser KJ, Venkatesh SK, Ben-Abraham EI, Ehman RL (2015) Feasibility of using 3D MR elastography to determine pancreatic stiffness in healthy volunteers. J Magn Reson Imaging 41:369–375. https://doi.org/10.1002/JMRI.24572
12. Tomalka A, Borsdorf M, Böl M, Siebert T (2017) Porcine stomach smooth muscle force depends on history-effects. Front Physiol 8:289206. https://doi.org/10.3389/FPHYS.2017.00802/BIBTEX
13. Shigeta Y, Saleh T, Benedetti G, Caciolli L, Chang J, Zambaiti E et al (2024) Stomach engineering: region-specific characterization of the decellularized porcine stomach. Pediatr Surg Int 40:1–9. https://doi.org/10.1007/S00383-023-05591-Y/FIGURES/4
14. Chanda A, Singh G (2023) Applications, challenges, and future opportunities. In: Materials horizons: from nature to nanomaterials, pp 85–92. https://doi.org/10.1007/978-981-99-2225-3_8/COVER
15. Gupta V, Chanda A (2022) Biomechanics of skin grafts: effect of pattern size, spacing and orientation. Eng Res Express 4:015006. https://doi.org/10.1088/2631-8695/AC48CB
16. Singh G, Chanda A (2022) Biomechanical modeling of progressive wound healing: a computational study. Biomed Eng Adv 4:100055. https://doi.org/10.1016/J.BEA.2022.100055
17. Egorov VI, Schastlivtsev IV, Prut EV, Baranov AO, Turusov RA (2002) Mechanical properties of the human gastrointestinal tract. J Biomech 35:1417–1425. https://doi.org/10.1016/S0021-9290(02)00084-2

18. Chung IK, Kim EJ, Lee MS, Kim HS, Park SH, Lee MH et al (2000) Bleeding Dieulafoy's lesions and the choice of endoscopic method: comparing the hemostatic efficacy of mechanical and injection methods. Gastrointest Endosc 52:721–724. https://doi.org/10.1067/MGE.2000.108040
19. Natali AN, Carniel EL, Gregersen H (2009) Biomechanical behaviour of oesophageal tissues: material and structural configuration, experimental data and constitutive analysis. Med Eng Phys 31:1056–1062. https://doi.org/10.1016/J.MEDENGPHY.2009.07.003
20. Leberfinger AN, Dinda S, Wu Y, Koduru SV, Ozbolat V, Ravnic DJ et al (2019) Bioprinting functional tissues. Acta Biomater 95:32–49. https://doi.org/10.1016/J.ACTBIO.2019.01.009
21. Gregersen H, Gilja OH, Hausken T, Heimdal A, Gao C, Matre K et al (2002) Mechanical properties in the human gastric antrum using B-mode ultrasonography and antral distension. Am J Physiol Gastrointest Liver Physiol 283. https://doi.org/10.1152/AJPGI.00144.2001/ASSET/IMAGES/LARGE/H30820931005.JPEG
22. Lim YJ, Deo D, Singh TP, Jones DB, De S (2009) In situ measurement and modeling of biomechanical response of human cadaveric soft tissues for physics-based surgical simulation. Surg Endosc 23:1298–1307. https://doi.org/10.1007/s00464-008-0154-z
23. Zhao J, Liao D, Gregersen H (2005) Tension and stress in the rat and rabbit stomach are location- and direction-dependent. Neurogastroenterol Motil 17:388–398. https://doi.org/10.1111/j.1365-2982.2004.00635.x
24. Zhao J, Liao D, Chen P, Kunwald P, Gregersen H (2008) Stomach stress and strain depend on location, direction and the layered structure. J Biomech 41:3441–3447. https://doi.org/10.1016/j.jbiomech.2008.09.008
25. Rosen J, Brown JD, De S, Sinanan M, Hannaford B (2008) Biomechanical properties of abdominal organs in vivo and postmortem under compression loads. J Biomech Eng 130. https://doi.org/10.1115/1.2898712
26. Gungor-Ozkerim PS, Inci I, Zhang YS, Khademhosseini A, Dokmeci MR (2018) Bioinks for 3D bioprinting: an overview. Biomater Sci 6:915–946. https://doi.org/10.1039/C7BM00765E
27. Vijayavenkataraman S, Yan WC, Lu WF, Wang CH, Fuh JYH (2018) 3D bioprinting of tissues and organs for regenerative medicine. Adv Drug Deliv Rev 132:296–332. https://doi.org/10.1016/J.ADDR.2018.07.004
28. Seidlits SK, Khaing ZZ, Petersen RR, Nickels JD, Vanscoy JE, Shear JB et al (2010) The effects of hyaluronic acid hydrogels with tunable mechanical properties on neural progenitor cell differentiation. Biomaterials 31:3930–3940. https://doi.org/10.1016/J.BIOMATERIALS.2010.01.125
29. Skardal A, Devarasetty M, Kang HW, Mead I, Bishop C, Shupe T et al (2015) A hydrogel bioink toolkit for mimicking native tissue biochemical and mechanical properties in bioprinted tissue constructs. Acta Biomater 25:24–34. https://doi.org/10.1016/J.ACTBIO.2015.07.030
30. Singh G, Gupta V, Chanda A (2022) Artificial skin with varying biomechanical properties. Mater Today Proc 62:3162–3166. https://doi.org/10.1016/J.MATPR.2022.03.433
31. Gupta V, Chanda A (2023) Expansion potential of novel skin grafts simulants with I-shaped auxetic incisions. Biomed Eng Adv 5:100071. https://doi.org/10.1016/J.BEA.2023.100071
32. Singh G, Chanda A (2023) Biofidelic gallbladder tissue surrogates. Adv Mater Process Technol. https://doi.org/10.1080/2374068X.2023.2198835
33. Chanda A, Singh G (2023) Tissues in functional organs—low stiffness. In: Materials horizons: from nature to nanomaterials, pp 33–48. https://doi.org/10.1007/978-981-99-2225-3_4/COVER
34. Gupta V, Singh G, Gupta S, Chanda A (2023) Expansion potential of auxetic prosthetic skin grafts: a review. Eng Res Express 5:022003. https://doi.org/10.1088/2631-8695/ACCFE5
35. Singh G, Chanda A (2023) Biofidelic tongue and tonsils tissue surrogates. In: Materials horizons: from nature to nanomaterials, Part F1471, pp 159–70. https://doi.org/10.1007/978-981-99-5064-5_10/COVER
36. Gupta V, Singh G, Chanda A (2023) High expansion auxetic skin graft simulants for severe burn injury mitigation. Eur Burn J 4:108–20. https://doi.org/10.3390/EBJ4010011

37. Chanda A (2018) Biomechanical modeling of human skin tissue surrogates. Biomimetics 3:18. https://doi.org/10.3390/BIOMIMETICS3030018
38. Singh G, Chanda A (2023) Development and biomechanical testing of artificial surrogates for vaginal tissue. Adv Mater Process Technol. https://doi.org/10.1080/2374068X.2023.2198837
39. Gupta V, Singh G, Chanda A (2023) Development of hierarchical auxetic skin graft simulants with high expansion potential. Biomed Eng Adv 5:100087. https://doi.org/10.1016/J.BEA.2023.100087
40. Singh G, Chanda A (2023) Development and biomechanical testing of human stomach tissue surrogates. In: Materials horizons: from nature to nanomaterials, Part F1471, pp 113–25. https://doi.org/10.1007/978-981-99-5064-5_7/COVER
41. Chanda A, Unnikrishnan V, Lackey K, Robbins J (2020) Biofidelic conductive soft tissue surrogates. Int J Polym Mater Polym Biomater 69:127–135. https://doi.org/10.1080/00914037.2018.1552856

Chapter 11
Gallbladder Tissue Simulants

The gallbladder is a small, pear-shaped organ located beneath the liver. It plays a role in the digestive system by storing and concentrating bile, a digestive fluid produced by the liver. When you eat, the gallbladder releases bile into the small intestine to help emulsify and break down fats. Gallbladder tissue is composed of various cell types and structures that contribute to its function. The inner lining of the gallbladder is made up of epithelial cells, which form a mucous membrane. These cells are responsible for the secretion of mucus and help prevent the bile from damaging the gallbladder walls. The gallbladder has a muscular wall made of smooth muscle tissue [1, 2]. This muscular layer contracts to expel bile into the small intestine when needed for digestion. The contraction is triggered by the hormone cholecystokinin (CCK), which is released in response to the presence of fats in the small intestine. Connective tissue provides structural support to the gallbladder and helps maintain its shape and integrity. Like any organ, the gallbladder is supplied with blood vessels and innervated by nerves. Blood vessels deliver nutrients and oxygen, while nerves help regulate the contraction and relaxation of the gallbladder muscles [3].

Gallbladder tissue can be subject to various conditions, including gallstones, inflammation (cholecystitis), and other diseases. In some cases, medical conditions may require the removal of the gallbladder, a procedure known as cholecystectomy. The absence of a gallbladder usually does not significantly impair digestion, as bile produced by the liver continues to be released into the small intestine, but it is no longer stored and concentrated in the gallbladder. The mechanical aspects of gallbladder tissue are crucial for its normal functioning in the digestive process [4]. The gallbladder is a muscular organ, and its mechanical features contribute to the storage and controlled release of bile. The gallbladder stores and concentrates bile produced by the liver. When food, especially fats, enters the small intestine, the gallbladder contracts forcefully to release the concentrated bile into the duodenum. This contraction is crucial for the efficient digestion and absorption of fats. The gallbladder has the ability to distend and expand to accommodate and store bile. This compliance allows it to store a concentrated form of bile, which is released in a

A. Chanda and G. Singh, *Soft Tissue Simulants*, Biomedical Materials for Multi-functional Applications, https://doi.org/10.1007/978-981-97-3060-5_11

controlled manner when needed for digestion. Distension also helps in preventing continuous and uncontrolled secretion of bile [5–7].

Understanding the mechanical aspects of gallbladder tissue is crucial for assessing its normal function and diagnosing issues such as gallstones or inflammation. Disorders affecting the mechanical properties of the gallbladder can impact its ability to store and release bile effectively, leading to digestive problems and other complications.

11.1 Traditional Gallbladder Tissues

Traditional gallbladder tissues typically refer to biological samples collected from the gallbladder for diagnostic or research purposes. Obtaining gallbladder tissues for research or diagnostic purposes typically involves the use of various sources, depending on the specific goals and requirements of the study. These tissues can be obtained through surgical procedures or post-mortem examinations (see Fig. 11.1). Gallbladder tissues are often obtained during surgical procedures, such as cholecystectomy (removal of the gallbladder). Surgeons collect tissue samples for examination, diagnosis, or research. Gallbladder tissues can be obtained during autopsies for post-mortem examination. This is a source of tissues for research or understanding the pathology of gallbladder diseases. Tissue banks and repositories store human tissues, including gallbladder tissues, for use in research. Researchers can request samples from these repositories for various studies. Established gallbladder cell lines, derived from primary gallbladder tissues, are available for in vitro studies. These cell lines provide a continuous and reproducible source of gallbladder cells for experiments. In preclinical studies, researchers may use gallbladder tissues obtained from animal models, such as mice, rats, or larger animals. Animal models can help simulate specific conditions or diseases [1, 3, 8]. It's important to note that ethical and legal considerations, including patient consent and adherence to privacy regulations, are crucial when obtaining and using gallbladder tissues for research or diagnostic purposes. Researchers should follow appropriate guidelines and institutional review board (IRB) approvals when working with human tissues.

11.1.1 Challenges with Traditional Gallbladder Tissues

While traditional gallbladder tissues have been valuable for diagnostic and research purposes, their use comes with several challenges. Obtaining human gallbladder tissues for research or diagnostic purposes is subject to the availability of suitable samples. Access to human tissue samples may be limited due to ethical considerations, donor availability, and the need for informed consent. Human gallbladder tissues can exhibit significant variability among individuals, making it challenging to establish standardized baselines for research or diagnostic purposes. Obtaining

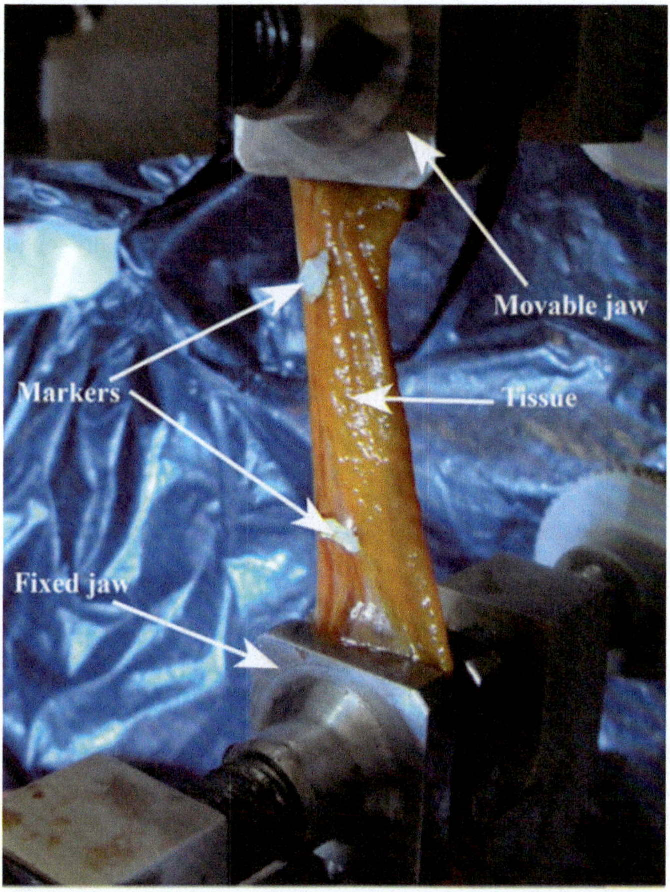

Fig. 11.1 Tensile testing of the human cadaveric gallbladder tissue [3]

informed consent for using human gallbladder tissues is essential, and ethical guidelines must be followed [1, 2]. This process can be logistically challenging and may limit the availability of samples. Ensuring the privacy and confidentiality of donors is crucial when working with human tissues, adding another layer of complexity to the ethical considerations. Human tissues are subject to degradation over time, even with proper preservation methods. This can impact the quality of samples and limit their usability for certain applications. Proper storage and transportation of human tissues require specialized facilities and procedures, adding logistical challenges to their use [9, 10].

Gallbladder diseases can manifest in various ways, and the heterogeneity of disease presentations can complicate the interpretation of results obtained from gallbladder tissues. Obtaining samples from individuals with specific gallbladder conditions may be challenging, limiting the ability to study rare or less common diseases.

Collecting, processing, and analyzing gallbladder tissues can be resource-intensive, requiring specialized equipment, skilled personnel, and financial investment. The cost and resource requirements associated with traditional gallbladder tissues have led to an increased interest in alternative methods, such as in vitro models and simulation techniques [11–13]. To address some of these challenges, researchers and clinicians have explored alternative approaches, including the use of gallbladder tissue simulants, in vitro models, and advanced imaging techniques. These alternatives aim to provide more standardized, reproducible, and ethically sound platforms for gallbladder-related studies.

11.2 Gallbladder Tissues Simulants

"Gallbladder tissue simulants" refer to materials or substances designed to mimic the properties of real gallbladder tissues for various purposes, including medical training, research, and device testing. These simulants aim to replicate the physical and mechanical characteristics of gallbladder tissue to provide a realistic environment without the ethical and logistical challenges associated with using real human or animal tissues. There are various factors affecting the tissue simulants including material composition, texture and elasticity, mechanical properties, visual appearance, thermal properties, end usage, etc. [13]. Various materials are widely used to replicate the soft, flexible nature of gallbladder tissues. The material composition is adjusted to replicate the texture and elasticity of real gallbladder tissue. This ensures that the simulant behaves realistically when manipulated or subjected to surgical instruments. Simulants often have a color and transparency similar to real gallbladder tissues. This visual realism enhances the training experience for medical professionals and researchers [14, 15].

11.2.1 Types of Gallbladder Tissue Simulants

Gallbladder tissue simulants are materials or substances designed to mimic the properties of real gallbladder tissues. These simulants serve various purposes, including medical training, research, and device testing. Several types of gallbladder tissue simulants exist, each tailored to specific applications. Silicone-based polymers are commonly used for creating soft tissue simulants. These materials can replicate the flexibility, elasticity, and tactile properties of real gallbladder tissue [4]. Silicone simulants are often used in surgical training models. Like silicone, polyurethane elastomers are flexible and can be formulated to simulate the mechanical properties of biological tissues. They are used in the development of tissue simulants for training and research purposes. Hydrogels made from agar or gelatin can mimic the softness and water content of real tissues. These materials are often used in the development of phantoms for ultrasound imaging and other diagnostic procedures. PVA hydrogels can be tailored to simulate the mechanical properties of gallbladder tissues. They are often used in the creation of tissue-mimicking phantoms for imaging studies [16–18].

Solid plastic models are sometimes used to represent gallbladder anatomy for educational purposes. These models may lack the softness and deformability of real tissues but can provide a static representation for anatomical reference. Rubber-based materials can be used to create gallbladder models with a degree of flexibility. These models are often employed for anatomical training purposes in medical education. Tissues from animals, such as pigs or cows, can be used to create biological tissue substitutes. While not synthetically engineered, these tissues may serve as alternatives for certain training and testing scenarios [19–21]. In the recent decade, 3D printing technology has allowed the creation of customizable gallbladder models with specific anatomical features. Various materials, including flexible filaments, can be used to simulate tissue properties [22]. Some 3D-printed models incorporate multiple materials to simulate different tissue layers and structures within the gallbladder, providing a more realistic representation. Hybrid models combine different materials to replicate the complex mechanical and structural properties of gallbladder tissues more accurately [23–25]. Choosing the appropriate type of gallbladder tissue simulant depends on the specific application, such as surgical training, medical device testing, or imaging studies. Each type of simulant has its advantages and limitations, and the selection is often based on the desired level of realism and functionality required for a particular use case.

11.2.2 Key Considerations in the Development of Gallbladder Tissue Simulant

The development of gallbladder tissue simulants involves careful consideration of various factors to ensure that the simulants accurately mimic the properties of real gallbladder tissues. Simulants should replicate the elasticity and compliance of gallbladder tissues, allowing them to deform and respond realistically to surgical instruments or external forces. Mimicking the distensibility of the gallbladder during the storage and release of bile is crucial for simulating physiological behavior. If applicable, the simulant should exhibit peristaltic movement to replicate the rhythmic contractions of the gallbladder. Simulants should have a color and transparency similar to real gallbladder tissues, enhancing visual realism for training and educational purposes [12, 13, 26]. Replicating internal structures such as blood vessels, ducts, and other anatomical features contributes to a more accurate representation of the gallbladder. If relevant to the application, simulants may be designed to respond to temperature changes similarly to real tissues, contributing to a more realistic training environment. Simulants used for ultrasound training or imaging studies should have acoustic properties similar to those of real gallbladder tissues. Simulants should be made from non-toxic materials to ensure the safety of users, especially in educational and training settings [1, 27, 28]. The ability to customize the properties of the simulant, such as elasticity or anatomical features, allows for tailoring the simulant to specific training or research needs. Reproducibility is crucial for maintaining consistency across different batches of simulants, ensuring reliability in training and research outcomes. Also, simulants should be resistant to degradation over time, ensuring their durability and longevity for repeated use. Simulants that replicate the tactile sensation of real tissues contribute to a more immersive and effective training experience. The cost of materials and production should be reasonable, especially for educational or training purposes where cost-effectiveness is crucial [14, 29–31].

Developing gallbladder tissue simulants involves an interdisciplinary approach, combining expertise in materials science, biomechanics, and medical education. Regular testing and validation against real tissues or imaging modalities are essential to ensure that the simulants effectively meet their intended purposes. Additionally, collaboration with medical professionals and educators can provide valuable insights into the specific requirements for training and educational simulations.

11.3 Applications of Gallbladder Tissue Simulants

Gallbladder tissue simulants find applications in various fields, contributing to research, medical training, device testing, and more. Simulants are used for hands-on training in gallbladder surgeries, such as laparoscopic cholecystectomy. They allow surgeons, residents, and medical students to practice incisions, dissection, and suturing techniques in a controlled and realistic environment [1, 4]. Gallbladder tissue simulants are particularly valuable for training in minimally invasive procedures. They help individuals develop skills in using laparoscopic instruments, trocars, and navigation within the abdominal cavity. Simulants accurately replicate the anatomy of the gallbladder, enabling students to study its structure, location, and relationships with surrounding organs. Simulants can also be designed to mimic pathological conditions for a more comprehensive educational experience [8]. Gallbladder tissue simulants serve as test models for evaluating the performance and safety of medical devices used in gallbladder surgeries. This includes testing laparoscopic instruments, imaging devices, and other surgical tools. Simulants are used in biomechanical studies to understand the mechanical properties of gallbladder tissues. This research contributes to the development of more realistic simulants and enhances our understanding of the gallbladder's behavior during various physiological processes [2, 6, 27]. Simulants can be tailored to mimic specific gallbladder diseases, facilitating research into the mechanisms and progression of conditions such as gallstones, cholecystitis, and gallbladder cancer [4].

Simulants offer a platform for surgeons to rehearse and plan gallbladder surgeries before performing them on actual patients. This can improve surgical outcomes and reduce the risk of complications. Simulant-based simulations provide a standardized way to assess the skills and proficiency of medical professionals in gallbladder-related procedures. Objective evaluations can be conducted to ensure competence. Simulants are used to train healthcare professionals to communicate effectively with patients regarding gallbladder-related procedures [32–35]. This includes explaining the surgical process, and potential risks, and obtaining informed consent. Simulants can be employed to visually demonstrate gallbladder conditions and surgical interventions to patients. This aids in patient education, helping them better understand their medical situation. Simulants provide a platform for surgeons to practice and refine new surgical techniques for gallbladder surgeries. This fosters innovation and continuous improvement in surgical practices [1, 2]. Surgeons can use simulants to advance their skills, explore novel approaches, and stay abreast of emerging technologies and methodologies in gallbladder surgery. Simulants address ethical concerns related to the use of human or animal tissues for training purposes. They provide an ethical alternative for training and research while maintaining high standards in medical education. In summary, gallbladder tissue simulants play a vital role in advancing medical education, supporting research endeavors, enhancing surgical skills, and contributing to the overall improvement of patient care in the field of gallbladder-related procedures [14].

11.4 Challenges and Future Direction for Gallbladder Tissue Simulants

11.4.1 Biomechanical Complexity

Challenge: Replicating the intricate biomechanical properties of the gallbladder, including its distensibility, elasticity, and peristaltic movement, presents a significant challenge. Achieving a realistic simulation of these dynamic behaviors is essential for effective training and research applications [13, 14].

Future Direction: Continued advancements in materials science and engineering may lead to the development of more sophisticated materials with enhanced biomechanical properties, providing a closer match to real gallbladder tissues [4, 15].

11.4.2 Tissue Heterogeneity

Challenge: The gallbladder can exhibit variations in tissue composition and properties among individuals, making it challenging to create a universal simulant that accurately represents this heterogeneity [13, 14].

Future Direction: Customizable simulants that allow adjustments to mimic specific patient characteristics may be a future direction. Incorporating imaging data from real patients could contribute to more personalized simulant development [4, 15].

11.4.3 Ultrasound Imaging Realism

Challenge: Achieving accurate ultrasound imaging realism is crucial for training and research. Simulants need to replicate the acoustic properties of real gallbladder tissues to ensure the effectiveness of ultrasound procedures [13, 14].

Future Direction: Advances in materials with tailored acoustic properties and the integration of realistic anatomical features in simulant designs may improve ultrasound imaging realism [4, 15].

11.4.4 Integration of Fluid Dynamics

Challenge: Simulating the behavior of fluids within the gallbladder, such as bile, is a complex task. The dynamic interaction of fluids with gallbladder tissues during contraction and relaxation is important for realistic simulations [13, 14].

Future Direction: Development of simulant models that incorporate fluid dynamics, potentially using innovative technologies such as microfluidics, to enhance the realism of gallbladder tissue simulants [4, 15].

11.4.5 Ethical Considerations

Challenge: Ensuring ethical considerations in simulant development, especially if human-derived materials or data are used, requires careful attention. Obtaining informed consent and addressing privacy concerns are critical [13, 14].

Future Direction: Clear guidelines and standards for ethical considerations in simulant development, along with increased transparency and communication with stakeholders, can help address these challenges [4, 15].

11.4.6 Validation and Standardization

Challenge: Establishing standardized methods for validating the effectiveness of gallbladder tissue simulants is a challenge. Without widely accepted benchmarks, comparing and assessing the realism of different simulants can be difficult [13, 14].

Future Direction: Collaboration among researchers, medical professionals, and educators to establish standardized validation protocols and criteria for gallbladder tissue simulants can enhance their credibility and effectiveness [4, 15].

In addition, several other factors should be emphasized for the better development and progress of gallbladder tissue simulants. First, integrating various simulation modalities, such as virtual reality, haptic feedback, and augmented reality, with physical simulators can enhance the overall training experience [6, 27]. This approach could provide a comprehensive simulation platform for gallbladder-related procedures. Second, advancements in medical imaging technologies and data processing may enable the creation of patient-specific gallbladder tissue simulants. This could facilitate personalized training scenarios and enhance the relevance of simulations to individual patient cases [4, 13, 33]. Third, the use of smart materials that can respond dynamically to external stimuli, such as temperature or pressure, could further enhance the realism of gallbladder tissue simulants. These materials may mimic the physiological responses of real tissues more accurately. Fourth, implementing a feedback loop involving end-users, such as medical students, residents, and practicing surgeons, can help identify areas for improvement in simulant design and functionality. This iterative process can lead to continuous refinement and innovation [14, 15, 36].

In conclusion, addressing the challenges and exploring future directions in gallbladder tissue simulant development require a multidisciplinary approach, technological innovation, and collaboration across various fields. The ongoing advancements in materials science, imaging technologies, and simulation methodologies hold promise for creating more realistic and effective gallbladder tissue simulants.

References

1. Li WG, Hill NA, Ogden RW, Smythe A, Majeed AW, Bird N et al (2013) Anisotropic behaviour of human gallbladder walls. J Mech Behav Biomed Mater 20:363–375. https://doi.org/10.1016/J.JMBBM.2013.02.015
2. Fridrichs J, Hamel B, Kelder W, van den Hoed E, van den Heuvel MC, Hulscher JBF et al (2023) Human precision-cut cystic duct and gallbladder slices: a novel method for studying cholangiopathies. Front Pediatr 11:1058319. https://doi.org/10.3389/FPED.2023.1058319/BIBTEX
3. Karimi A, Shojaei A, Tehrani P (2017) Measurement of the mechanical properties of the human gallbladder. J Med Eng Technol 41:541–545. https://doi.org/10.1080/03091902.2017.1366561
4. Singh G, Chanda A (2021) Mechanical properties of whole-body soft human tissues: a review. Biomed Mater 16:062004. https://doi.org/10.1088/1748-605X/AC2B7A
5. Li WG, Luo XY, Hill NA, Ogden RW, Smythe A, Majeed A et al (2011) A mechanical model for CCK-induced acalculous gallbladder pain. Ann Biomed Eng 39:786–800. https://doi.org/10.1007/S10439-010-0205-1/FIGURES/9
6. Li WG, Luo XY, Hill NA, Smythe A, Chin SB, Johnson AG et al (2008) Correlation of mechanical factors and gallbladder pain. Comput Math Methods Med 9:27–45. https://doi.org/10.1080/17486700701780266
7. Liao D, Duch BU, Stødkilde-Jørgensen H, Zeng YJ, Gregersen H, Kassab GS (2004) Tension and stress calculations in a 3-D Fourier model of gall bladder geometry obtained from MR images. Ann Biomed Eng 32:744–755. https://doi.org/10.1023/B:ABME.0000030239.46843.01/METRICS

8. Li WG, Luo XY, Hill NA, Ogden RW, Smythe A, Majeed AW et al (2012) A quasi-nonlinear analysis of the anisotropic behaviour of human gallbladder wall. J Biomech Eng 134. https://doi.org/10.1115/1.4007633/450903
9. Genovese K, Casaletto L, Humphrey JD, Lu J (2014) Digital image correlation-based point-wise inverse characterization of heterogeneous material properties of gallbladder in vitro. Proc R Soc A Math Phys Eng Sci 470. https://doi.org/10.1098/RSPA.2014.0152
10. Poston GJ, Singh P, Maclellan DG, Yao CZ, Uchida T, Townsend CM et al (1988) Age-related changes in gallbladder contractility and gallbladder cholecystokinin receptor population in the guinea pig. Mech Ageing Dev 46:225–236. https://doi.org/10.1016/0047-6374(88)90127-3
11. Gupta V, Singh G, Chanda A (2023) Development of hierarchical auxetic skin graft simulants with high expansion potential. Biomed Eng Adv 5:100087. https://doi.org/10.1016/J.BEA.2023.100087
12. Gupta V, Singla R, Singh G, Chanda A (2023) Development of soft composite based anisotropic synthetic skin for biomechanical testing. Fibers 11:55. https://doi.org/10.3390/FIB11060055
13. Singh G, Chanda A (2023) Biofidelic gallbladder tissue surrogates. Adv Mater Process Technol. https://doi.org/10.1080/2374068X.2023.2198835
14. Chanda A, Singh G (2023) Applications, challenges, and future opportunities. In: Materials horizons: from nature to nanomaterials, pp 85–92. https://doi.org/10.1007/978-981-99-2225-3_8/COVER
15. Chanda A, Singh G (2023) Introduction to human tissues. In: Materials horizons: from nature to nanomaterials, pp 1–12. https://doi.org/10.1007/978-981-99-2225-3_1/COVER
16. Mondal S, Das S, Nandi AK (2020) A review on recent advances in polymer and peptide hydrogels. Soft Matter 16:1404–1454. https://doi.org/10.1039/c9sm02127b
17. Talebian S, Mehrali M, Taebnia N, Pennisi CP, Kadumudi FB, Foroughi J et al (2019) Self-healing hydrogels: the next paradigm shift in tissue engineering? Adv Sci 6. https://doi.org/10.1002/advs.201801664
18. Wang Y, Adokoh CK, Narain R (2018) Recent development and biomedical applications of self-healing hydrogels. Expert Opin Drug Deliv 15:77–91. https://doi.org/10.1080/17425247.2017.1360865
19. Lowry OH, Gilligan DR, Katersky EM (1941) The determination of collagen and elastin in tissues, with results obtained in various normal tissues from different species. J Biol Chem 139:795–804. https://doi.org/10.1016/s0021-9258(18)72951-7
20. Neuman RE, Logan MA (1950) The determination of collagen and elastin in tissues. J Biol Chem 186:549–556. https://doi.org/10.1016/s0021-9258(18)56248-7
21. Sharabi M, Mandelberg Y, Benayahu D, Benayahu Y, Azem A, Haj-Ali R (2014) A new class of bio-composite materials of unique collagen fibers. J Mech Behav Biomed Mater 36:71–81. https://doi.org/10.1016/J.JMBBM.2014.04.008
22. Chhikara K, Singh G, Gupta S, Chanda A (2022) Progress of additive manufacturing in fabrication of foot orthoses for diabetic patients: a review. Ann 3D Print Med 8:100085. https://doi.org/10.1016/J.STLM.2022.100085
23. Leberfinger AN, Dinda S, Wu Y, Koduru SV, Ozbolat V, Ravnic DJ et al (2019) Bioprinting functional tissues. Acta Biomater 95:32–49. https://doi.org/10.1016/J.ACTBIO.2019.01.009
24. Jammalamadaka U, Tappa K (2018) Recent advances in biomaterials for 3D printing and tissue engineering. J Funct Biomater 9:22. https://doi.org/10.3390/JFB9010022
25. Matai I, Kaur G, Seyedsalehi A, McClinton A, Laurencin CT (2020) Progress in 3D bioprinting technology for tissue/organ regenerative engineering. Biomaterials 226:119536. https://doi.org/10.1016/J.BIOMATERIALS.2019.119536
26. Gupta V, Gupta S, Chanda A (2023) Development of an ultra-low-cost planar biaxial tester for soft tissue characterization. Biomed Phys Eng Express 9:025011. https://doi.org/10.1088/2057-1976/ACB940
27. Haller W, Sharif K, Millar AJW, Brown RM, McKiernan PJ (2010) Gallbladder dysfunction in cholesterol ester storage disease. J Pediatr Gastroenterol Nutr 50:555–558. https://doi.org/10.1097/MPG.0B013E31819F659B

28. Hinescu ME, Ardeleanu C, Gherghiceanu M, Popescu LM (2007) Interstitial Cajal-like cells in human gallbladder. J Mol Histol 38:275–284. https://doi.org/10.1007/S10735-007-9099-0/FIGURES/6
29. Stevenson TC, Whitlock JA, Martonick N, Cheatham SW, Reeves A, McGowan C et al (2023) Descriptive analysis of forces applied by trained clinicians during 2-handed instrument-assisted soft tissue mobilization. J Athl Train 58:60–64. https://doi.org/10.4085/1062-6050-282-21
30. Chanda A, McClain S (2019) Mechanical modeling of healthy and diseased calcaneal fat pad surrogates. Biomimetics 4:1. https://doi.org/10.3390/BIOMIMETICS4010001
31. Gupta V, Chanda A (2022) Biomechanics of skin grafts: effect of pattern size, spacing and orientation. Eng Res Express 4:015006. https://doi.org/10.1088/2631-8695/AC48CB
32. Makode S, Singh G, Chanda A (2021) Development of novel anisotropic skin simulants. Phys Scr 96:125019. https://doi.org/10.1088/1402-4896/AC2EFD
33. Singh G, Gupta V, Chanda A (2022) Artificial skin with varying biomechanical properties. Mater Today Proc 62:3162–3166. https://doi.org/10.1016/J.MATPR.2022.03.433
34. Gupta V, Singh G, Chanda A (2022) Development and testing of skin grafts models with varying slit orientations. Mater Today Proc 62:3462–3467. https://doi.org/10.1016/J.MATPR.2022.04.282
35. Singh G, Chanda A (2022) Biomechanical modeling of progressive wound healing: a computational study. Biomed Eng Adv 4:100055. https://doi.org/10.1016/J.BEA.2022.100055
36. Chanda A, Singh G (2023) Tissues in functional organs—low stiffness. In: Materials horizons: from nature to nanomaterials, pp 33–48. https://doi.org/10.1007/978-981-99-2225-3_4/COVER

Chapter 12
Simulants for Arteries

Arterial tissue refers to the tissues that make up arteries, which are blood vessels responsible for carrying oxygenated blood away from the heart to various parts of the body. Arteries play a crucial role in the circulatory system, and their structure is specialized to withstand the high pressure and pulsatile flow of blood that occurs as a result of the heart's pumping action. Arterial tissue has three layers, i.e., tunica intima, tunica media, and tunica adventitia (see Fig. 12.1). Tunica intima is the innermost layer of arterial tissue and consists of two components known as endothelium and subendothelial layer. The endothelium is a single layer of endothelial cells that form a smooth, thin lining on the interior surface of the artery [1, 2]. This endothelial layer is in direct contact with the blood, promoting smooth blood flow and preventing clotting. The subendothelial layer is a layer of connective tissue that provides support to the endothelium. The middle layer of arterial tissue is tunica media and has smooth muscle cells and elastic fibers [1–3]. Smooth muscle cells constitute the majority of the tunica media and are arranged in concentric layers. The contraction and relaxation of smooth muscle cells regulate the diameter of the artery, influencing blood flow and blood pressure. Interspersed among smooth muscle cells, elastic fibers provide elasticity to the artery, allowing it to stretch during systole (heart contraction) and recoil during diastole (heart relaxation). This property helps maintain steady blood flow. The outermost layer is primarily composed of connective tissue and is known as tunica adventitia (or tunica externa). Predominantly collagen fibers, providing structural support and protecting the artery. Small blood vessels within the adventitia supply nutrients and oxygen to the outer layers of the artery [4].

A. Chanda and G. Singh, *Soft Tissue Simulants*, Biomedical Materials for Multi-functional Applications, https://doi.org/10.1007/978-981-97-3060-5_12

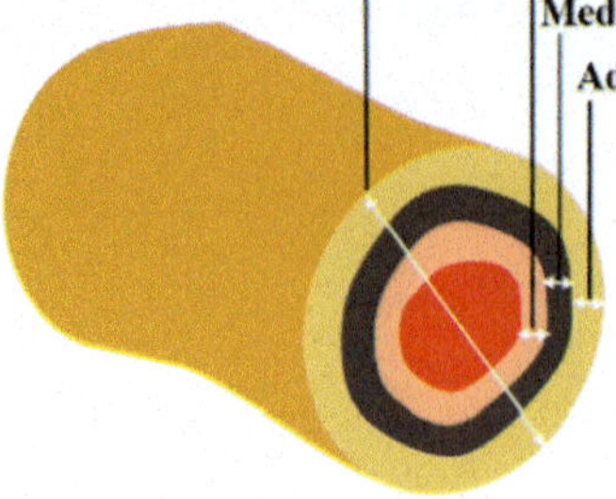

Fig. 12.1 Different layers of a coronary artery with dimensions [5]

12.1 Types and Functions of Arteries

The structure of arterial tissue is designed to maintain the integrity and function of arteries under varying conditions. Arteries are classified into three main types based on their size and function. Elastic arteries are large arteries near the heart, such as the aorta. Rich in elastic fibers, allowing them to stretch and recoil to accommodate the surge of blood during systole (heart contraction) and maintain pressure during diastole (heart relaxation). The second type of arteries are muscular arteries [6, 7]. These medium-sized arteries that distribute blood to various organs and tissues have a well-developed tunica media with smooth muscle to regulate blood flow. Arterioles, the third type, are small arteries that connect to capillaries. They have a thinner structure compared to larger arteries but still possess the three layers of arterial tissue. Arterioles play a crucial role in regulating blood flow into capillary beds and controlling blood pressure [2].

Arteries transport oxygenated blood away from the heart to supply various tissues and organs throughout the body. The tunica media, with its smooth muscle and elastic fibers, allows arteries to regulate blood pressure by adjusting the diameter of the vessel. Arterial tissue, especially muscular arteries and arterioles, plays a key role in distributing blood to specific regions based on the body's physiological needs [8, 9]. Understanding the structure and function of arterial tissue is essential in the context of cardiovascular health. Changes in arterial structure, such as atherosclerosis (accumulation of plaque), can lead to various cardiovascular diseases, including hypertension, coronary artery disease, and stroke. Researchers and healthcare professionals study arterial tissue to gain insights into these conditions and develop effective diagnostic and therapeutic strategies [10].

12.2 Significance of Arterial Tissue

The significance of arterial tissue from a mechanical aspect is crucial in understanding how arteries function under mechanical forces and stresses, and how alterations in mechanical properties can contribute to various cardiovascular diseases. Arterial tissue must withstand the pulsatile flow of blood and the pressure generated by the heart's pumping action. Arterial tissue must exhibit compliance and elasticity to accommodate the rhythmic changes in blood pressure during the cardiac cycle [7, 11, 12]. The tunica media is rich in elastic fibers, allowing arteries to stretch during systole and recoil during diastole. This elasticity helps maintain a steady flow of blood and prevents excessive pressure fluctuations. The smooth muscle cells in the tunica media contract and relax, regulating the diameter of the artery. Contraction (vasoconstriction) and relaxation (vasodilation) of smooth muscle influence the resistance to blood flow and, consequently, blood pressure. This mechanical control is vital for blood pressure regulation [13]. The mechanical properties of arterial tissue determine the speed at which the pulse wave travels through the arteries. Changes in arterial stiffness can affect the timing and reflection of the pulse wave, influencing blood flow patterns and potentially contributing to cardiovascular complications.

12.2.1 Biomechanics of Arterial Tissue

Arterial tissue experiences mechanical stress due to blood pressure. Understanding the biomechanics of the arterial wall helps assess factors such as wall stress, which is critical in evaluating the risk of aneurysm formation or rupture. Arterial tissue exhibits viscoelastic behavior, meaning it has both elastic and viscous properties. This enables the artery to absorb and dissipate energy, contributing to its ability to withstand mechanical forces. Arterial tissue can undergo remodeling in response to changes in mechanical forces. For example, in hypertension, arteries may thicken to adapt to increased pressure [14, 15]. Chronic changes in mechanical forces can lead to structural alterations in arterial tissue, contributing to conditions such as atherosclerosis and arterial stiffness. Arterial stiffness is considered a biomarker for cardiovascular health. Increased stiffness is associated with aging and various cardiovascular risk factors [4, 16]. Understanding the mechanical properties of arterial tissue is essential in interpreting diagnostic tests such as arterial imaging, Doppler ultrasound, and pulse wave analysis. This aids in predicting how arteries respond to different mechanical loads and understanding the impact of structural changes. In summary, the mechanical significance of arterial tissue is pivotal in maintaining proper cardiovascular function. Alterations in the mechanical properties of arteries can contribute to the development and progression of cardiovascular diseases, making it an essential area of study in both clinical and research settings [4, 17].

12.3 Arterial Tissues Simulants

Arterial tissue simulants are synthetic materials or models designed to mimic the mechanical, structural, and physiological properties of real arterial tissues. These simulants are used in various medical and research applications for testing, training, and development purposes. Arterial tissue simulants are employed to assess the performance and safety of medical devices like stents and grafts [17–19]. This helps in optimizing the design and materials used in these devices. Simulants are used to simulate arterial conditions for testing catheters and balloons used in angioplasty and other vascular interventions. Arterial tissue simulants provide a realistic training environment for vascular surgeons to practice procedures like anastomosis, bypass grafting, and endarterectomy. Simulants are used for training in minimally invasive procedures such as endovascular stent placement and angiography. Researchers use arterial tissue simulants to study the biomechanical properties of arteries, such as elasticity, compliance, and stress distribution. This information is valuable for understanding how arteries respond to various forces. Simulants are incorporated into computational models for finite element analysis, allowing researchers to simulate and analyze the behavior of arterial tissues under different conditions. Simulants are used to calibrate and test medical imaging devices like ultrasound, CT scanners, and MRI machines to ensure accurate and reliable results [15, 20]. Radiologists use arterial tissue simulants to enhance their skills in interpreting imaging results and diagnosing vascular conditions. Simulants are employed to create models of atherosclerosis, aiding researchers in studying the progression and characteristics of this common vascular disease. Arterial tissue simulants are used in pharmacological research to assess the effects of drugs on the vascular system [4, 21].

12.3.1 Characteristics of Arterial Tissue Simulants

Simulants are designed to replicate the mechanical behavior of arterial tissues, including elasticity, compliance, and the ability to withstand physiological pressures. The composition of arterial tissue simulants often includes synthetic materials such as silicone rubber, polyurethane, or other polymers [22–24]. These materials are chosen for their durability, flexibility, and biocompatibility. Simulants aim to replicate the anatomical features of arteries, including vessel diameter, wall thickness, and branching patterns. Some arterial tissue simulants are designed to simulate blood flow through arteries, allowing researchers to study hemodynamics and fluid dynamics in a controlled environment [10, 17].

12.3.2 Challenges and Considerations

Arterial tissues in the human body exhibit considerable biological variability. Creating a simulant that accurately represents this variability can be challenging. Simulants should be durable enough to withstand repeated use in training and testing scenarios without significant degradation. While arterial tissue simulants are not meant for implantation in the body, ensuring biocompatibility is important, especially when used in close proximity to biological tissues [10, 13, 25]. In summary, arterial tissue simulants play a crucial role in medical training, research, and device development by providing a controlled and reproducible environment to simulate the properties and behaviors of real arterial tissues. The development and use of these simulants contribute to advancements in vascular medicine and surgery [4].

12.4 Types of Arterial Tissue Simulants

Arterial tissue simulants come in various types, each designed to serve specific purposes in medical training, device testing, and research. The choice of simulant depends on the intended application and the properties required to accurately replicate arterial tissue characteristics. Some common types of arterial tissue simulants are silicone rubber simulants, polyurethane simulants, elastomeric materials, hydrogel simulants, combination models, 3D-printed simulants, and bioengineered tissue models [26–28]. Silicone rubber is a common material for arterial tissue simulants due to its flexibility, durability, and ease of molding. Silicone rubber simulants are often used for training in vascular surgery, particularly for practicing anastomosis, suturing, and other surgical techniques (see Fig. 12.2). Polyurethane is another synthetic material employed for arterial tissue simulants. It can be formulated to replicate the mechanical properties of real arteries. Polyurethane simulants are used for testing and evaluating medical devices such as stents, grafts, and catheters. They are also suitable for biomechanical studies. Elastomers, including materials like latex or rubber-based compounds, are used to create arterial tissue simulants with stretchable and elastic properties. These simulants are often used to replicate the elasticity of arterial walls, making them suitable for biomechanical studies and simulations of arterial deformation [29]. Hydrogels are water-absorbing polymers that can be tailored to mimic the mechanical properties of soft tissues, including arteries. Hydrogel simulants are utilized in applications where a more tissue-like feel and response to surgical instruments are required, such as in surgical training and procedural simulations [4, 30].

Some arterial tissue simulants use a combination of materials to replicate different layers of arterial walls accurately. These models aim to provide a more comprehensive representation of arterial tissue, making them suitable for various applications, including medical device testing and training. Advances in 3D printing technology

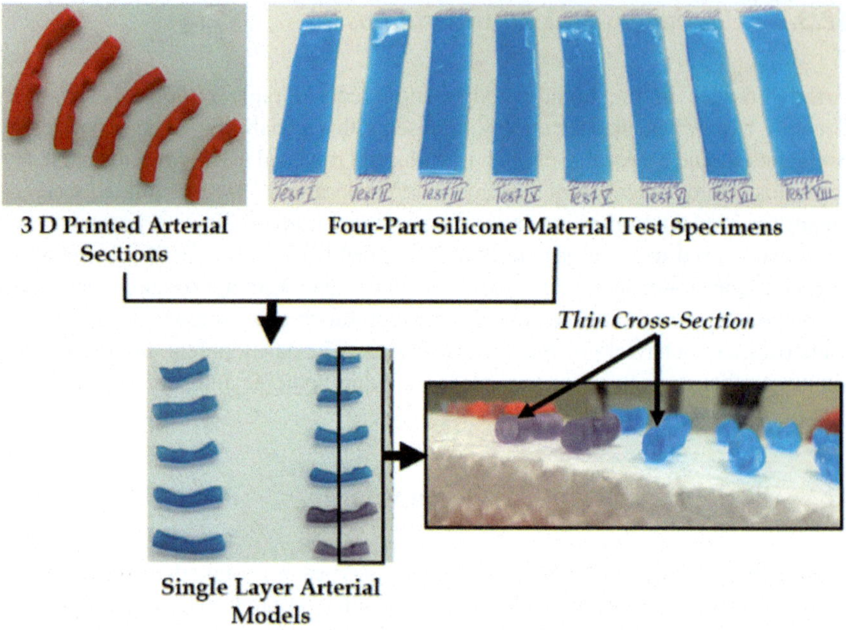

Fig. 12.2 Fabrication of single-layer (intimal) coronary artery models [5]

allow the creation of customized arterial tissue simulants using a variety of materials, including thermoplastics and bioresorbable polymers. 3D-printed arterial tissue simulants can replicate complex anatomical structures and are often used for surgical training and preoperative planning [22, 31, 32] (see Fig. 12.3). In some advanced applications, bioengineered tissue constructs incorporating biological and synthetic components are used to create arterial tissue simulants. These constructs may allow for a more realistic representation of tissue response and interactions, particularly in regenerative medicine and tissue engineering research.

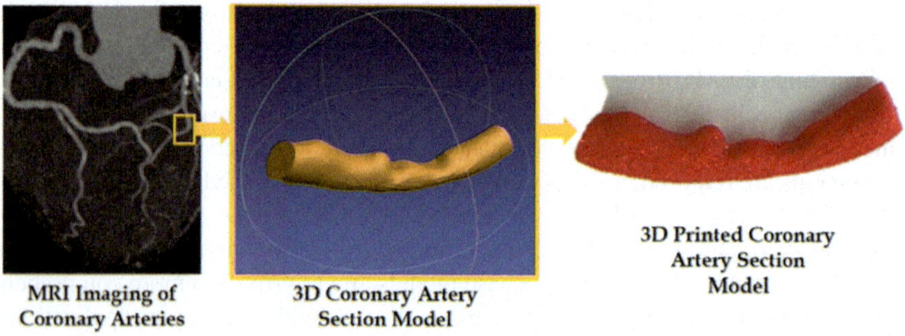

Fig. 12.3 Development of 3D-printed arterial model from MRI images of coronary artery [5]

12.4.1 *Considerations*

While arterial tissue simulants are not meant for implantation, ensuring biocompatibility is important for applications where the simulant comes into contact with biological tissues or fluids. Second, simulants should strive to replicate the anatomical features of arteries, including vessel geometry, wall thickness, and branching patterns. Third, simulants should be durable enough to withstand repeated use, especially in training scenarios [10, 17, 33]. The choice of arterial tissue simulant depends on the specific goals of the application, ranging from training surgeons to testing medical devices and conducting biomechanical studies. Advances in materials science and technology continue to contribute to the development of more realistic and versatile arterial tissue simulants [4].

12.5 Why Arterial Tissue Simulants Are in Trend?

Arterial tissue simulants have gained popularity and are considered advantageous in various fields due to several reasons. The increasing trend in using arterial tissue simulants is driven by their versatility, cost-effectiveness, and ability to address specific challenges in medical training, research, and device development. Obtaining human arterial tissues for research and training can raise ethical concerns, especially when involving invasive procedures or post-mortem tissues [7, 20]. Arterial tissue simulants provide an ethical alternative, allowing researchers and medical professionals to conduct studies and training without the need for human specimens. Traditional methods of obtaining arterial tissues, such as cadaveric specimens, can be expensive to acquire and maintain. Arterial tissue simulants are often more cost-effective and provide a reusable option for various applications, contributing to reduced overall costs in research and medical training [4]. Arterial tissue simulants can be precisely engineered to replicate specific mechanical, anatomical, and physiological properties. These customized models can closely mimic the conditions of interest, enhancing reproducibility and allowing for consistent experimental conditions (see Fig. 12.4).

Arterial tissue simulants offer controlled environments for biomechanical studies, allowing researchers to investigate the behavior of arterial tissues under specific mechanical loads. These studies contribute to a deeper understanding of arterial mechanics, aiding in the development of medical devices and interventions. Arterial tissue simulants provide a realistic and controlled environment for medical training, allowing healthcare professionals to practice procedures and enhance their skills. Such detailed arterial tissue simulants are used in testing and optimizing medical devices, such as stents and grafts, for safety and efficacy. Device developers can

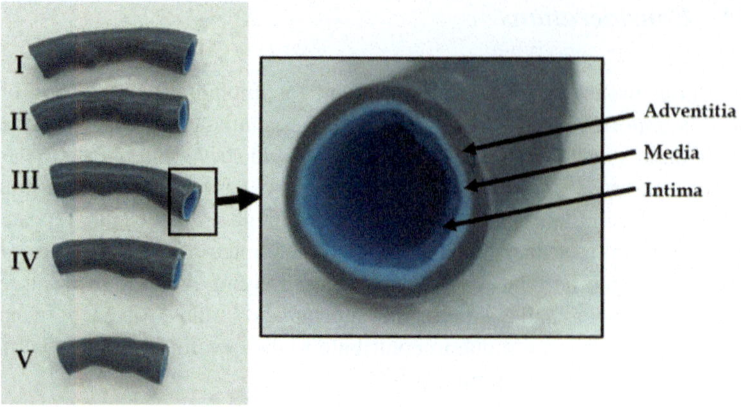

Fig. 12.4 Three-layer arterial models [5]

perform extensive testing in controlled conditions, speeding up the design and optimization process before clinical trials. Arterial tissue simulants contribute to standardizing research methodologies by providing consistent and controlled experimental conditions. Standardization facilitates more accurate comparisons across studies, promoting the reproducibility of research findings [4, 17].

In summary, the increasing trend in using arterial tissue simulants is driven by their ethical considerations, cost-effectiveness, customization capabilities, and their contributions to advancements in medical training, research, and device development. As technology and materials science continue to advance, arterial tissue simulants are likely to play an even more prominent role in medical education and research.

12.6 Arterial Tissue for Various Applications

Arterial tissue, comprising the walls of arteries in the circulatory system, is crucial for various applications in medical research, diagnostics, and therapeutic interventions. Understanding the properties and functions of arterial tissue is essential for addressing cardiovascular diseases and advancing medical treatments.

12.6.1 Cardiovascular Research

Arterial tissue is used to investigate the development and progression of atherosclerosis, a condition characterized by the buildup of plaque within the arteries. Researchers study arterial tissue to understand the mechanisms involved and to develop potential interventions. Arterial tissue serves as a fundamental model for studying vascular biology, including the interaction between endothelial cells, smooth muscle cells, and other components of the arterial wall [4].

12.6.2 Device Testing and Development

Arterial tissue is utilized to test the effectiveness and safety of stents and grafts. These devices are designed to treat conditions such as atherosclerosis and aneurysms by providing structural support or replacing damaged sections of arteries. Engineers and researchers study the mechanical properties of arterial tissue to design and optimize medical devices, ensuring they can withstand physiological stresses without causing damage [4].

12.6.3 Drug Development

Arterial tissue is used to assess the effects of drugs on blood vessels and vascular function. This is crucial for developing medications that target conditions like hypertension, atherosclerosis, and vascular inflammation. Understanding the behavior of arterial tissue helps in designing drug delivery systems that can target specific regions of the vasculature [4].

12.6.4 Diagnostic Imaging

Arterial tissue is imaged using angiography techniques to visualize the structure and function of arteries. This is crucial for diagnosing conditions such as stenosis, aneurysms, and arterial blockages. Arterial tissue is employed in studies using ultrasound and magnetic resonance imaging (MRI) to assess blood flow, detect abnormalities, and evaluate vascular health [4].

12.6.5 Biomedical Engineering

Arterial tissue is used in biomechanical models to simulate blood flow, pressure, and vessel wall behavior. These models aid in understanding the mechanical forces acting on arteries and can contribute to the design of medical devices. Arterial tissue is investigated in the field of tissue engineering to develop artificial blood vessels or enhance the repair of damaged arteries using bioengineered constructs [4].

12.6.6 Training and Education

Arterial tissue models are utilized for training surgeons, allowing them to practice procedures such as arterial bypass grafting and angioplasty in a controlled environment before performing surgeries on patients. Arterial tissue is used in educational settings to teach medical students and healthcare professionals about the anatomy, physiology, and pathology of arteries [4].

12.6.7 Regenerative Medicine

Arterial tissue is explored in regenerative medicine for cell-based therapies. Researchers investigate the potential of stem cells to repair damaged arteries and promote tissue regeneration. Arterial tissue can guide the development of biomaterials and scaffolds for tissue engineering applications aimed at repairing or replacing damaged arteries [4].

Understanding arterial tissue and its applications is integral to advancing cardiovascular medicine, improving diagnostics, and developing innovative treatments for a range of vascular conditions. Researchers and healthcare professionals continue to explore new avenues to leverage arterial tissue in various fields of medicine and biomedical research.

References

1. Urakami-Harasawa L, Shimokawa H, Nakashima M, Egashira K, Takeshita A (1997) Importance of endothelium-derived hyperpolarizing factor in human arteries. J Clin Invest 100:2793–2799. https://doi.org/10.1172/JCI119826
2. McKee JA, Banik SSR, Boyer MJ, Hamad NM, Lawson JH, Niklason LE et al (2003) Human arteries engineered in vitro. EMBO Rep 4:633–638. https://doi.org/10.1038/SJ.EMBOR.EMBOR847/ASSET/42E70E24-2CC9-4709-972D-5CE9579921E9/ASSETS/GRAPHIC/EMBR847-FIG-0003-M.JPG
3. Fleischer S, Naveed Tavakol D, Vunjak-Novakovic G, Fleischer S, Tavakol DN, Vunjak-Novakovic G (2020) From arteries to capillaries: approaches to engineering human vasculature. Adv Funct Mater 30:1910811. https://doi.org/10.1002/ADFM.201910811
4. Singh G, Chanda A (2021) Mechanical properties of whole-body soft human tissues: a review. Biomed Mater 16:062004. https://doi.org/10.1088/1748-605X/AC2B7A
5. Chanda A, Curry K (2018) Patient-specific biofidelic human coronary artery surrogates. J Mech Med Biol 18. https://doi.org/10.1142/S0219519418500495
6. Ferrara A, Pandolfi A (2008) Numerical modelling of fracture in human arteries. Comput Methods Biomech Biomed Eng 11:553–567. https://doi.org/10.1080/10255840701771743
7. Claes E, Atienza JM, Guinea GV, Rojo FJ, Bernal JM, Revuelta JM et al (2010) Mechanical properties of human coronary arteries. In: 2010 annual international conference of the IEEE engineering in medicine and biology society, EMBC'10, pp 3792–3795. https://doi.org/10.1109/IEMBS.2010.5627560

8. Hutchins GM, Kessler-Hanna A, Moore GW (1988) Development of the coronary arteries in the embryonic human heart. Circulation 77:1250–1257. https://doi.org/10.1161/01.CIR.77.6.1250
9. Friedman MH, Hutchins GM, Brent Bargeron C, Deters OJ, Mark FF (1981) Correlation between intimal thickness and fluid shear in human arteries. Atherosclerosis 39:425–436. https://doi.org/10.1016/0021-9150(81)90027-7
10. Chanda A, Singh G (2023) Applications, challenges, and future opportunities. In: Materials horizons: from nature to nanomaterials, pp 85–92. https://doi.org/10.1007/978-981-99-2225-3_8/COVER
11. Herrera J, Henke CA, Bitterman PB (2018) Extracellular matrix as a driver of progressive fibrosis. J Clin Invest 128:45–53. https://doi.org/10.1172/JCI93557
12. Mendelson K, Schoen FJ (2006) Heart valve tissue engineering: concepts, approaches, progress, and challenges. Ann Biomed Eng 34:1799–1819. https://doi.org/10.1007/S10439-006-9163-Z/FIGURES/5
13. Johnston BM, Johnston PR, Corney S, Kilpatrick D (2004) Non-Newtonian blood flow in human right coronary arteries: steady state simulations. J Biomech 37:709–720. https://doi.org/10.1016/J.JBIOMECH.2003.09.016
14. Jensen BC, Swigart PM, Laden ME, DeMarco T, Hoopes C, Simpson PC (2009) The alpha-1D is the predominant alpha-1-adrenergic receptor subtype in human epicardial coronary arteries. J Am Coll Cardiol 54:1137–1145. https://doi.org/10.1016/J.JACC.2009.05.056
15. Holzapfel GA, Ogden RW, Holzapfel GA, Ogden RW (2010) Constitutive modelling of arteries. Proc R Soc A Math Phys Eng Sci 466:1551–1597. https://doi.org/10.1098/RSPA.2010.0058
16. Vijayavenkataraman S, Yan WC, Lu WF, Wang CH, Fuh JYH (2018) 3D bioprinting of tissues and organs for regenerative medicine. Adv Drug Deliv Rev 132:296–332. https://doi.org/10.1016/J.ADDR.2018.07.004
17. Chanda A, Singh G (2023) Introduction to human tissues. In: Materials horizons: from nature to nanomaterials, pp 1–12. https://doi.org/10.1007/978-981-99-2225-3_1/COVER
18. Singh G, Gupta V, Chanda A (2022) Artificial skin with varying biomechanical properties. Mater Today Proc 62:3162–3166. https://doi.org/10.1016/J.MATPR.2022.03.433
19. Singh G, Chanda A (2023) Biofidelic gallbladder tissue surrogates. Adv Mater Process Technol. https://doi.org/10.1080/2374068X.2023.2198835
20. Figueroa CA, Vignon-Clementel IE, Jansen KE, Hughes TJR, Taylor CA (2006) A coupled momentum method for modeling blood flow in three-dimensional deformable arteries. Comput Methods Appl Mech Eng 195:5685–5706. https://doi.org/10.1016/J.CMA.2005.11.011
21. Hayashi K, Handa H, Nagasawa S, Okumura A, Moritake K (1980) Stiffness and elastic behavior of human intracranial and extracranial arteries. J Biomech 13:175–184. https://doi.org/10.1016/0021-9290(80)90191-8
22. Gupta V, Singla R, Singh G, Chanda A (2023) Development of soft composite based anisotropic synthetic skin for biomechanical testing. Fibers 11:55. https://doi.org/10.3390/FIB11060055
23. Gupta V, Singh G, Chanda A (2023) High expansion auxetic skin graft simulants for severe burn injury mitigation. Eur Burn J 4:108–120. https://doi.org/10.3390/EBJ4010011
24. Gupta V, Singh G, Chanda A (2023) Modeling of metamaterial based incision patterns for generating high expansions in skin grafts. Clin Biomech 110:106118. https://doi.org/10.1016/J.CLINBIOMECH.2023.106118
25. Vignon-Clementel IE, Alberto Figueroa C, Jansen KE, Taylor CA (2006) Outflow boundary conditions for three-dimensional finite element modeling of blood flow and pressure in arteries. Comput Methods Appl Mech Eng 195:3776–3796. https://doi.org/10.1016/J.CMA.2005.04.014
26. Singh G, Chanda A (2023) Biofidelic tongue and tonsils tissue surrogates. In: Materials horizons: from nature to nanomaterials, Part F1471, pp 159–170. https://doi.org/10.1007/978-981-99-5064-5_10/COVER
27. Gupta V, Singh G, Chanda A (2022) Development and testing of skin grafts models with varying slit orientations. Mater Today Proc 62:3462–3467. https://doi.org/10.1016/J.MATPR.2022.04.282

28. Singh G, Chanda A (2023) Development and biomechanical testing of human stomach tissue surrogates. In: Materials horizons: from nature to nanomaterials, Part F1471, pp 113–125. https://doi.org/10.1007/978-981-99-5064-5_7/COVER
29. Martins PALS, Natal Jorge RM, Ferreira AJM (2006) A comparative study of several material models for prediction of hyperelastic properties: application to silicone-rubber and soft tissues. Strain 42:135–147. https://doi.org/10.1111/j.1475-1305.2006.00257.x
30. Vikatmaa P, Juutilainen V, Kuukasjärvi P, Malmivaara A (2008) Negative pressure wound therapy: a systematic review on effectiveness and safety. Eur J Vasc Endovasc Surg 36:438–448. https://doi.org/10.1016/J.EJVS.2008.06.010
31. Mu J, Sun T, Leung CLA, Oliveira JP, Wu Y, Wang H et al (2023) Application of electrochemical polishing in surface treatment of additively manufactured structures: a review. Prog Mater Sci 136:101109. https://doi.org/10.1016/J.PMATSCI.2023.101109
32. Khan A ur R, Huang K, Khalaji MS, Yu F, Xie X, Zhu T et al (2021) Multifunctional bioactive core-shell electrospun membrane capable to terminate inflammatory cycle and promote angiogenesis in diabetic wound. Bioact Mater 6:2783–800. https://doi.org/10.1016/J.BIOACTMAT.2021.01.040
33. Chanda A, Singh G (2023) Tissues in functional organs—low stiffness. In: Materials horizons: from nature to nanomaterials, pp 33–48. https://doi.org/10.1007/978-981-99-2225-3_4/COVER

Chapter 13
Tissue Simulants for Ballistics Testing

Ballistic testing of tissues is a scientific and forensic method used to understand the effects of projectiles, such as bullets or projectiles fired from firearms, on biological tissues. This type of testing is crucial for forensic investigations, medical research, and the development of protective gear. Some key aspects of ballistic testing of tissues include understanding the wound patterns and potential causes of injuries in cases involving firearms [1]. First, to study the effects of bullets or projectiles on human or animal tissues for medical and surgical advancements. Second, to assess the performance of protective gear, such as body armor, against ballistic threats. In recent scenarios, tissue simulants are materials that are extensively used to mimic the mechanical and ballistic characteristics of human tissues [2, 3]. However, human or animal cadavers, as well as tissue simulants, are used widely for testing purposes. Such tissue simulants help in reconstructing shooting incidents and determining the range and angle of fire. For the applications in medical advances, they provide insights into the treatment of gunshot wounds and the development of improved medical procedures [4].

13.1 State of the Art of Ballistic Tissue Simulants

13.1.1 Cadavers

The majority of studies in the field of ballistics are conducted to examine ballistic wounds or the potential for bullets to cause wounds on the different tissues or organs of the body. Accordingly, such tissues/organs that are involved ought to keep their regular or natural biomechanical qualities, specifically their tensile strength. As a result, tissues/organs taken from recently deceased bodies have been used [5, 6]. In a similar manner, tissues/organs acquired from preserved cadavers without the use of embalming have also been used in ballistic wound studies. In the course of this study,

A. Chanda and G. Singh, *Soft Tissue Simulants*, Biomedical Materials for Multi-functional Applications, https://doi.org/10.1007/978-981-97-3060-5_13

contradictory conclusions have been revealed about the biomechanical characteristics of organs and tissues, specifically the tensile strength of these structures. There was no influence on the biomechanical parameters, but there were significant changes in modulus of elasticity and tensile strength [7–9]. These results varied from that to a decrease in tensile strength. A connection has been made between the latter alterations and the dehydration that occurs during freezing. Hemolysis, extracellular fractures, and loss of cellular fluid are some of the histological alterations that may be seen in tissues that have been frozen and then defrosted [10, 11]. In most cases, cadavers that are utilized for study come from programs that donate their bodies. As a result, they are often from persons of different ages which results in the variation of mechanical properties (such as elastic modulus) across them. Additionally, such tissue samples give rise to other ethical concerns when taking these factors into consideration. Thus, cadaveric tissue samples should be avoided for the ballistic testing research, and searching for other alternatives may be more beneficial [12–14].

13.1.2 Animal Models

Ballistic testing using animal models has been a controversial and ethically sensitive topic. The use of animals in such experiments raises ethical concerns regarding animal welfare and has led to increased efforts to develop alternative methods. Many countries and institutions have regulations and guidelines that restrict or prohibit the use of animals in ballistic testing. There have been instances where animals, particularly pigs and goats, were used in ballistic testing to simulate the effects of gunshot wounds on human tissue [15, 16]. These studies aimed to understand the mechanisms of injury and develop protective measures for military and law enforcement personnel. Humans have been reported to physically share comparable anatomy with pigs, which has led to the consideration of porcine tissue as a natural human tissue-mimicking material. Pigs, however, have more thicker skin tissue than humans; as a result, it is possible that pigs are not a suitable alternative for humans in laboratory experiments [17, 18]. In recent years, there has been a growing emphasis on using alternative methods such as computer simulations, synthetic models, and cadavers to replace or reduce the use of animals in ballistic testing. These alternatives aim to provide accurate and relevant data without causing harm to live animals.

13.1.3 Ballistic Ordnance Gelatin

Ordnance gelatine is derived from collagens found in animal-based items using hot water acidic extraction. It is offered in regularities ranging from fifty to three hundred Bloom, which is a measure of the gelatine's firmness and stiffness [17, 19]. The Bloom Number is not the only factor that determines the strength and stiffness of the material; temperature and concentration in the preparation process also have a role. Gelatine-based ballistic ordnance is now recognized as a human soft tissue-mimicking material, which when calibrated appropriately as per standards may be utilized in ballistic tests. Appropriate calibration did not start until the middle of the 1980s when it was discovered that gelatine that had not been calibrated included defects [12, 20]. Several experiments were carried out to determine the degree to which these simulants resembled the tissues of porcine animals. It was discovered that it is capable of imitating the biomechanical qualities of the skin layers, fats, muscles, and fascia that are present in a pig thigh. Researchers have come to the conclusion that the attributes of projectile-based tissue penetrations are the same as that of human soft tissues. This is due to the fact that pig tissues are comparable to human tissues. As a result of its ability to reproduce elasticity that is comparable to that of some human tissues, it is helpful because it enables the viewing and photographic portrayal of the wound profiles caused by projectiles [21]. Due to the fact that it does not possess the biomechanical qualities of other organs and tissues, it is possible that it may not correctly reflect all of the soft tissues that are found in the human body. Furthermore, when a bullet penetrates gelatin, radial fractures appear. This is a distinct behavior from the way human tissues react when they are pierced by a bullet [7, 20]. Radial cracks are different from the behavior of human tissues. Because of this, it could be difficult to translate the data that were obtained in ballistics ordinance gelatine to actual wounding that takes place in the human body. Despite the fact that ballistic ordnance gelatine makes it possible to investigate general tissue injuries caused by projectiles, it only mimics pig tissues, and more specifically porcine muscle [13, 15, 22–24]. As a result, it does not precisely duplicate the biomechanical qualities and reactions of the many organs and tissues that are found in the human body.

13.1.4 Synthetic Tissue Simulants

When doing research on wound ballistics, it is a standard practice to make use of synthetic tissue simulants. Compared to biological tissues, the tissue simulants are more attractive due to limited biosafety issues, infection possibilities, and the ability to conduct high-speed photography for damage analysis [18, 25]. The problems that are arising with ballistic ordnance gelatine are being addressed by the development of new homogenous synthetic simulants. The transparent synthetic medium known as Perma-Gel is stable at room temperature and has qualities that are comparable to those of 10% ordnance gelatine when it is at 4 degrees Celsius. They are gaining recognition, notably among scholars in the United States of America [25], and it is possible to reassemble them. Physically Associating Gelatines (PAG) are also being produced in the United States. These gelatines are created from triblock copolymers, which display deformation behaviors that are comparable to those of ballistics ordnance gelatine. As a result, PAG is an alternate tissue simulant. A translucent gel candle is another excellent soft tissue simulant. It is made with kraton and its combination with paraffin oil (white) and can comply with the Internal Revenue Service-based set calibration requirements. Due to the fact that it is transparent, this has the added benefit of enabling superior imaging of both the permanent and temporary cavities. Furthermore, in contrast to ordnance gelatine, this substance is not impacted by bacterial contamination [26]. On the other hand, these newly developed simulants are still homogenous and are not capable of describing all of the injuries that occur internally and in organs.

13.1.5 Models Based on Anatomy

The process of mimicking the properties of human tissues and organs is a challenging task. In the plethora of cases, including the analysis of motor vehicle collisions, the testing of projectiles, and the creation of body armor, a significant amount of research has been undertaken and is now being conducted in order to construct anatomical models. The production of synthetic bone, blood, organs, skin, and muscles is required in order to simulate the human body. This kind of simulation may ultimately result in the development of computational Finite Element Analysis (FEA) models (see Fig. 13.1) [12, 18, 27].

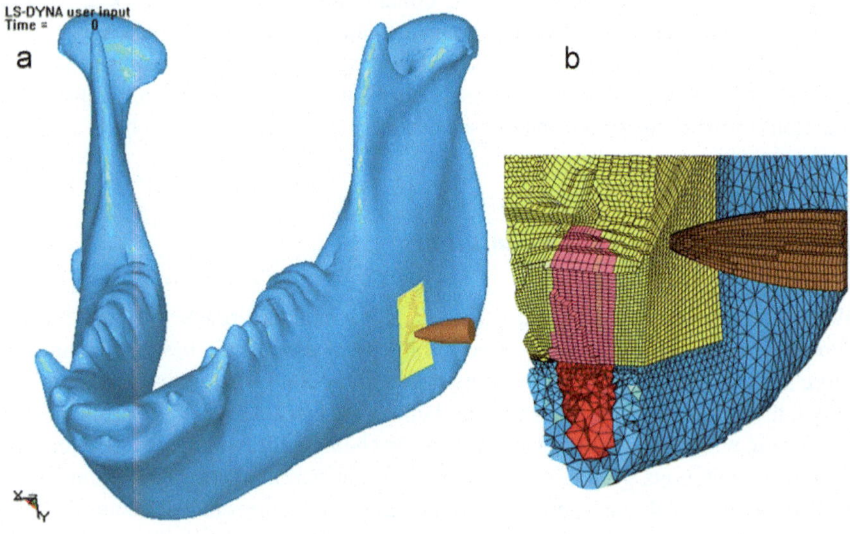

Fig. 13.1 The finite element model (a) and localized section (b) of the impact area meshing [28]

13.2 Ballistic Simulants for Head-Neck Evaluation

13.2.1 Brain Simulants

To create a realistic simulant that can imitate the human brain when it is subjected to high-velocity impact, it is essential to have a solid understanding of the mechanical characteristics and anatomy of the human brain. Additionally, the brain is a viscoelastic substance, which means that it has both viscous and elastic qualities. This is the case with the majority of biological tissues. The strain rate and the amount of time both have an effect on its mechanical moduli. Increasing the strain rate as well as the strain range causes the material to become more rigid. It is also possible that temperature has a substantial impact on the mechanical characteristics of the brain tissue, as shown by these findings [18, 29, 30].

In the past, brains from animal and cadaveric models were utilized to simulate and investigate the effects of gunshot wounds on human beings [31]. As a result of the implementation of ethical norms, limits have been put into place, which restrict the use of both living animals and cadavers. A skull structure that is sufficiently comparable to that of a human being is not found in any animal that is readily accessible. At the moment, a significant number of forensic investigators are using gelatin that is ten percent as a substitute for brain tissue. Recent research [32] has indicated that 10% gelatin composition was less accurate than 3% gelatin variant in simulating the brain tissue. Collagen is an animal tissue-based protein which is the component that makes up gelatin. Gelatin, when combined with water, forms an amorphous polymeric gel having structural properties widely distinct from that of brain tissue by virtue of its composition. Varying degrees of viscoelastic and sometimes plastic material properties are shown by soft materials. These behaviors are determined by the composition of the material. The behavior of gelatin in ballistic testing is characterized by a high degree of elasticity, and it does not bend plastically as the brain does. This was shown by Lazarjan et al. [32] by the use of low-speed photography capturing an air rifle (0.22 inch caliber) projectile shot into varying gelatin compositions and compared with fresh bovine brain tissue. With this investigation, gelatin was found to exhibit greater elasticity than brain tissues. Furthermore, since gelatin is not fibrous and isotropic, its biomechanics and fracture properties are different from that of the brain [30, 33]. Therefore, it is not reasonable to anticipate that gelatin and the brain would behave similarly in ballistic analysis. The experimental testing setup of the gas gun system for projectile simulation is shown in Fig. 13.2.

The most recent research conducted by Falland-Cheung and colleagues [31] examined the characteristics of a variety of simulants and compared them to deer brain. The researchers also evaluated mixes of agar and glycerol as well as agar, glycerol, and water that were struck with air rifle-based projectiles. In both the damage patterns observed before and after the impacts, conditioned simulants exhibited behavior that was comparable to that of the brain of a deer. During the process of developing a "skin-skull-brain model," Thali et al. [35] employed at 4 degrees Celsius temperature-controlled gelatine 10% to serve as a representation of the brain. The scientists found

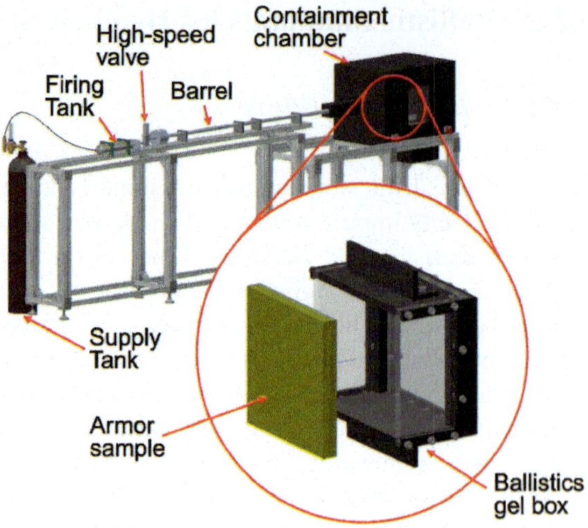

Fig. 13.2 Gas gun setup for the projectile simulation [34]

that the experimental bullet-based damage to the model and in realistic injuries were comparable. In recent studies, 10% gelatine-filled synthetic skulls were observed to showcase accurate damage patterns in ballistic tests [18, 36]. When the 10% gelatine was compared with the 3, 5, and 7% gelatine, as well as the PermagelTM, there was no statistically significant difference found. When discussing the characteristics that tissue simulants should have, Jussila [37] said that it is not necessary for them to have the exact same biomechanical properties as actual tissue. This is granted that "the results can be measured and appropriately extrapolated or scaled."

13.2.2 Skull Simulants

Studies on human skull-based ballistic tests, in both the presence or absence of a brain model, are limited [12, 15, 18]. In one of the earlier studies by Miller in 1970, skull penetration with metal-based projectiles was attempted. Watkins et al. [38] released some early work on dry testing of skulls filled with different compositions of gelatine. Nine "heads" were tested with ball bearings with a diameter of one-fourth of an inch (6.35 mm) and a mass of approximately 1 g traveling at 370–1000 m/s speeds. The velocities were measured during the impact and exit. Approximately a 370 m per second impact was able to just penetrate the entire skull, resulting in the deposition of 72 J; impacts at approximately 750 m per second deposited 250 J and caused "significant" damage; impacts at approximately 1,000 m per second deposited 440 J and caused "very severe" skull damage and fracture [39] [40].

13.2.3 Skin Simulants

Jussila et al. [3] reviewed the mechanical and ballistic properties of human skin simulants thoroughly using the available literature as their source material. They observed that the structural layers of human skin each possess unique characteristics and absorb varied quantities of impact energy. Furthermore, they observed that these characteristics alter depending on the location of the body as well as the age of the individual. They then proceeded to evaluate a number of synthetic and natural materials by comparing them to the values that were published for cadavers. During the measurements, both the elongation at break and the threshold velocity that a particular projectile must have in order to pierce the materials were taken into consideration. An upholstery "crust" cowhide with a semi-finished chrome tanned finish was found to be the most effective natural simulant. For the purpose of wound ballistic study, the authors noted that a synthetic material that was both user-friendly and high in fidelity was required. In addition, Falland-Cheung et al. [31] have discussed the ways in which the mechanical characteristics of human skin are influenced by aspects such as age, gender, and health, as well as the ways in which a dependable synthetic equivalent might be beneficial for impact testing. It was determined that dental silicones were superior to pig skin in terms of their mechanical qualities. Despite the fact that the characteristics of the pig skin and silicones were different, the strength of the silicone was comparable to human skin [18, 27, 30, 41–43].

The skin is stretched, partly crushed, and eventually ruptured as a result of an impacting bullet, which creates the opportunity for the bullet to penetrate the hypodermis layer. When searching for a skin simulant that is acceptable for wound ballistic testing, it is essential to take into consideration the threshold velocity v_{th} that a particular projectile must have in order to penetrate the skin. In turn, v_{th} is reliant on the mechanical qualities of the skin when it comes to the skin. The tensile strength of a material and its elongation at break (maximum elongation) are two of the many ways that the mechanical qualities of materials may be defined. However, these two characteristics seem to be the most significant. There is a correlation between the age of a person and the position of their skin on their body. The establishment of a skin simulant consequently necessitates the classification of a target individual. The definition of the reference person has been determined to be a guy who is thirty years old, according to the standpoint of law enforcement. The anterior thorax is the site that is considered to be the target skin in theoretical terms [15, 18, 22].

A number of researchers have isolated the skin that was taken from human cadavers and then placed it under the microscope [21, 25, 31]. Several studies have arrived at the conclusion that the sectional density "S" of the projectile is the primary factor that determines skin penetrability. This is calculated by dividing the mass of the bullet by its cross-sectional area in grams per square centimeter. They have also proposed a threshold for this.

To conclude, a good tissue simulant should have mechanical and penetration properties similar to real tissue in high-strain environments and must be repeatable and reproducible [1, 31, 36, 39].

References

1. Carr D, Lindstrom AC, Jareborg A, Champion S, Waddell N, Miller D et al (2015) Development of a skull/brain model for military wound ballistics studies. Int J Legal Med 129:505–510. https://doi.org/10.1007/S00414-014-1073-2/FIGURES/3
2. Casem DT, Dwivedi AK, Mrozek RA, Lenhart JL (2014) Compression response of a thermoplastic elastomer gel tissue surrogate over a range of strain-rates. Int J Solids Struct 51:2037–2046. https://doi.org/10.1016/J.IJSOLSTR.2013.12.028
3. Jussila J, Leppäniemi A, Paronen M, Kulomäki E (2005) Ballistic skin simulant. Forensic Sci Int 150:63–71. https://doi.org/10.1016/J.FORSCIINT.2004.06.039
4. Bracq A, Haugou G, Delille R, Lauro F, Roth S, Mauzac O (2017) Experimental study of the strain rate dependence of a synthetic gel for ballistic blunt trauma assessment. J Mech Behav Biomed Mater 72:138–147. https://doi.org/10.1016/J.JMBBM.2017.04.027
5. Maiden NR, Byard RW (2016) Unpredictable tensile strength biomechanics may limit thawed cadaver use for simulant research. Aust J Forensic Sci 48:54–58. https://doi.org/10.1080/00450618.2015.1025842
6. Bir C, Viano D, King A (2004) Development of biomechanical response corridors of the thorax to blunt ballistic impacts. J Biomech 37:73–79. https://doi.org/10.1016/S0021-9290(03)00238-0
7. Venkatasubramanian RT, Wolkers WF, Shenoi MM, Barocas VH, Lafontaine D, Soule CL et al (2010) Freeze-thaw induced biomechanical changes in arteries: Role of collagen matrix and smooth muscle cells. Ann Biomed Eng 38:694–706. https://doi.org/10.1007/S10439-010-9921-9/FIGURES/7
8. Venkatasubramanian RT, Grassl ED, Barocas VH, Lafontaine D, Bischof JC (2006) Effects of freezing and cryopreservation on the mechanical properties of arteries. Ann Biomed Eng 34:823–832. https://doi.org/10.1007/S10439-005-9044-X/FIGURES/10
9. Ming-Che W, Pins GD, Silver FH (1994) Collagen fibres with improved strength for the repair of soft tissue injuries. Biomaterials 15:507–512. https://doi.org/10.1016/0142-9612(94)90016-7
10. Epstein E, Munderloh N (1975) Isolation and characterization of CNBr peptides of human (alpha 1 (III))3 collagen and tissue distribution of (alpha 1 (I))2 alpha 2 and (alpha 1 (III))3 collagens. J Biol Chem 250:9304–9312. https://doi.org/10.1016/s0021-9258(19)40644-3
11. Neuman RE, Logan MA (1950) The determination of collagen and elastin in tissues. J Biol Chem 186:549–556. https://doi.org/10.1016/s0021-9258(18)56248-7
12. Chanda A, Singh G (2023) Applications, challenges, and future opportunities. Mater Horiz Nat Nanomater 85–92. https://doi.org/10.1007/978-981-99-2225-3_8/COVER
13. Singh G, Chanda A (2023) Biofidelic gallbladder tissue surrogates. Adv Mater Process Technol. https://doi.org/10.1080/2374068X.2023.2198835
14. Singh G, Chanda A (2023) Biofidelic tongue and tonsils tissue surrogates. Mater Horiz Nat Nanomater Part F1471:159–170. https://doi.org/10.1007/978-981-99-5064-5_10/COVER
15. Chanda A, Singh G (2023) Introduction to human tissues. Mater Horiz Nat Nanomater 1–12. https://doi.org/10.1007/978-981-99-2225-3_1/COVER
16. Gupta V, Singh G, Gupta S, Chanda A (2023) Expansion potential of auxetic prosthetic skin grafts: a review. Eng Res Express 5:022003. https://doi.org/10.1088/2631-8695/ACCFE5
17. Cronin DS (2011) Ballistic gelatin characterization and constitutive modeling. Conf Proc Soc Exp Mech Ser 1:51–55. https://doi.org/10.1007/978-1-4614-0216-9_7/COVER
18. Singh G, Chanda A (2021) Mechanical properties of whole-body soft human tissues: a review. Biomed Mater 16:062004. https://doi.org/10.1088/1748-605X/AC2B7A
19. Haag H-G, Herrmann C, Kampf W, Nordsiek K-H, Streck R, Zerpner D (1989) Functionalized polybutadiene oils as adhesion promotors for mineral fillers in rubber compounds. Die Angew Makromol Chemie 171:1–19. https://doi.org/10.1002/APMC.1989.051710101
20. Pandey PK, Harmukh A, Khan MK, Iqbal MA, Ganpule SG (2023) Ballistic response of skin simulant against fragment simulating projectiles. Def Technol 30:70–82. https://doi.org/10.1016/J.DT.2023.04.009

21. Mabbott A, Carr DJ, Champion S, Malbon C (2016) Comparison of porcine thorax to gelatine blocks for wound ballistics studies. Int J Legal Med 130:1353–1362. https://doi.org/10.1007/S00414-015-1309-9/TABLES/3
22. Singh G, Gupta V, Chanda A (2022) Artificial skin with varying biomechanical properties. Mater Today Proc 62:3162–3166. https://doi.org/10.1016/J.MATPR.2022.03.433
23. Singh G, Chanda A (2023) Development and biomechanical testing of artificial surrogates for vaginal tissue. Adv Mater Process Technol. https://doi.org/10.1080/2374068X.2023.2198837
24. Singh G, Chanda A (2023) Development and biomechanical testing of human stomach tissue surrogates. Mater Horiz Nat Nanomater Part F1471:113–125. https://doi.org/10.1007/978-981-99-5064-5_7/COVER
25. Appleby-Thomas GJ, Wood DC, Hameed A, Painter J, Le-Seelleur V, Fitzmaurice BC (2016) Investigation of the high-strain rate (shock and ballistic) response of the elastomeric tissue simulant Perma-Gel®. Int J Impact Eng 94:74–82. https://doi.org/10.1016/J.IJIMPENG.2016.04.003
26. Jussila J (2005) Measurement of kinetic energy dissipation with gelatine fissure formation with special reference to gelatine validation. Forensic Sci Int 150:53–62. https://doi.org/10.1016/j.forsciint.2004.06.038
27. Makode S, Singh G, Chanda A (2021) Development of novel anisotropic skin simulants. Phys Scr 96:125019. https://doi.org/10.1088/1402-4896/AC2EFD
28. Chen Y, Miao Y, Xu C, Zhang G, Lei T, Tan Y (2010) Wound ballistics of the pig mandibular angle: a preliminary finite element analysis and experimental study. J Biomech 43:1131–1137. https://doi.org/10.1016/J.JBIOMECH.2009.12.009
29. Chanda A, Singh G (2023) Tissues in functional organs—low stiffness. Mater Horiz Nat Nanomater 33–48. https://doi.org/10.1007/978-981-99-2225-3_4/COVER
30. Singh G, Chanda A (2023) Development and mechanical characterization of artificial surrogates for brain tissues. Biomed Eng Adv 5:100084. https://doi.org/10.1016/J.BEA.2023.100084
31. Falland-Cheung L, Piccione N, Zhao T, Lazarjan MS, Hanlin S, Jermy M et al (2016) Investigation of dental alginate and agar impression materials as a brain simulant for ballistic testing. Forensic Sci Int 263:169–175. https://doi.org/10.1016/J.FORSCIINT.2016.04.005
32. Lazarjan MS, Geoghegan PH, Jermy MC, Taylor M (2014) Experimental investigation of the mechanical properties of brain simulants used for cranial gunshot simulation. Forensic Sci Int 239:73–78. https://doi.org/10.1016/J.FORSCIINT.2014.03.022
33. Chanda A, Callaway C (2018) Tissue anisotropy modeling using soft composite materials. Appl Bionics Biomech 2018. https://doi.org/10.1155/2018/4838157
34. Goode T, Shoemaker G, Schultz S, Peters K, Pankow M (2019) Soft body armor time-dependent back face deformation (BFD) with ballistics gel backing. Compos Struct 220:687–698. https://doi.org/10.1016/J.COMPSTRUCT.2019.04.025
35. Thali MJ, Kneubuehl BP, Zollinger U, Dirnhofer R (2002) The, "Skin–skull–brain model": a new instrument for the study of gunshot effects. Forensic Sci Int 125:178–189. https://doi.org/10.1016/S0379-0738(01)00637-5
36. Mahoney PF, Carr DJ, Delaney RJ, Hunt N, Harrison S, Breeze J et al (2017) Does preliminary optimisation of an anatomically correct skull-brain model using simple simulants produce clinically realistic ballistic injury fracture patterns? Int J Legal Med 131:1043–1053. https://doi.org/10.1007/S00414-017-1557-Y/TABLES/7
37. Jussila J (2004) Preparing ballistic gelatine—review and proposal for a standard method. Forensic Sci Int 141:91–98. https://doi.org/10.1016/J.FORSCIINT.2003.11.036
38. Watkins FP, Pearce BP, Stainer MC (1988) Physical effects of the penetration of head simulants by steel spheres. J Trauma Inj Infect Crit Care 28:S40–54
39. De Boer HH (Hans), Van der Merwe AE (Lida), Soerdjbalie-Maikoe V (Vidija) (2016) Human cranial vault thickness in a contemporary sample of 1097 autopsy cases: relation to body weight, stature, age, sex and ancestry. Int J Legal Med 130:1371–1377. https://doi.org/10.1007/S00414-016-1324-5/TABLES/4
40. Smith MJ, Brickley MB, Leach SL (2007) Experimental evidence for lithic projectile injuries: improving identification of an under-recognised phenomenon. J Archaeol Sci 34:540–553. https://doi.org/10.1016/J.JAS.2006.06.008

41. Gupta V, Singh G, Chanda A (2023) Development of hierarchical auxetic skin graft simulants with high expansion potential. Biomed Eng Adv 5:100087. https://doi.org/10.1016/J.BEA.2023.100087
42. Gupta V, Singla R, Singh G, Chanda A (2023) Development of soft composite based anisotropic synthetic skin for biomechanical testing. Fibers 11:55. https://doi.org/10.3390/FIB11060055
43. Gupta V, Singh G, Chanda A (2023) High expansion auxetic skin graft simulants for severe burn injury mitigation. Eur Burn J 4:108–120. https://doi.org/10.3390/EBJ4010011

The manufacturer's authorised representative in the EU is Springer Nature Customer Service Centre GmbH, Europaplatz 3, 69115 Heidelberg, Germany. If you have any concerns regarding our products, please contact ProductSafety@springernature.com

Printed and bound by CPI Group (UK) Ltd, Croydon, CR0 4YY
01/10/2026
02225877-0008